UPCO

Physical Setting

PHYSICS

Kristofer M. Gigante
Physics Teacher
Guilderland Central High School
Guilderland Center, New York

UPCO-United Publishing Co., Inc.
21A Railroad Avenue
Albany, New York 12205
(518) 438-1600

CONTRIBUTING EDITORS/REVIEWERS

Julie R. Drahushuk
Physics Teacher

William J. Tucci
Mathematics and Science Section Chief
North Carolina Department of Public Instruction
(Retired)

Valley Central High School
Montgomery, NY (Retired)

ISBN 978-0-937323-26-7

1 2 3 4 5 6 7 8 9 0

CONTENTS

Introduction

"All science is either physics or stamp collecting." - Ernest Rutherford

Welcome to Physics and the understanding of the physical world! This book is broken into four main parts, which cover all of the basics for a high school course in Physics.

Part I: Mechanics includes the mathematics of Physics, the study of motion (kinematics), the study of the causes of motion (dynamics and forces), two-dimensional motion, momentum, and the study of mechanical energy.

Part II: Waves, Sound, and Light will explore the properties of mechanical waves, the study of sound, and light behavior. Reflection, refraction, diffraction, and interference of light waves are also discussed.

Part III: Electricity and Magnetism covers static electricity, electric fields, current electricity, and electromagnetism.

Part IV: Modern and Subatomic Physics introduces the reader to the world on a subatomic scale. This section includes the structure of the atom and its particles, development of the atomic model, quantum theory, and nuclear physics.

Wherever possible, there are connections to measurements made in a typical laboratory experiment, connections to the everyday world around us, and the notable scientists who have helped get us to the level of understanding we have today in Physics.

Finally, there are selected review questions embedded throughout the book for practice, a glossary of important terms, and practice Regents Exams in Physics so that the student can be well-prepared for the Physical Setting / Physics Regents Examination.

How Well Do You Know Your Physics Course?

The following questions and statements represent a general overview of the Regents Physics Course. This is not an all-inclusive list, but more of a starting point to diagnose the students' strengths and weaknesses. As an exercise, students should read each question or statement below and attempt as complete of an answer for each as possible. For each question or statement that is not clearly answered by the student, more information can be found in this book and/or by asking your teacher.

1. What are the fundamental units of measure?
2. What is a derived unit? Give some examples.
3. How do you convert from one unit of measure to another?
4. What does the graph of a direct relationship look like?
5. What does the graph of an inverse relationship look like? Inverse square?
6. What is the slope of a displacement vs. time graph?
7. What is the slope of a velocity vs. time graph?
8. How do you determine displacement from a velocity vs. time graph?
9. What is the numerical value of the slope of any horizontal line graph?
10. Can any motion graph have a vertical line? Why/Why not?
11. If an object has a positive velocity and a positive acceleration, describe the motion of the object.
12. If an object has a positive velocity and a negative acceleration, describe the motion of the object.
13. If an object has a negative velocity and a positive acceleration, describe the motion of the object.
14. If an object has a negative velocity and a negative acceleration, describe the motion of the object.
15. What is the difference between a vector and a scalar quantity?
16. Identify the important vector and scalar quantities in physics.
17. What is a vector resultant? How is it determined?
18. What must be the angle between vectors to give the maximum resultant?
19. What must be the angle between vectors to give the minimum resultant?
20. What is equilibrium in terms of forces?
21. If there is no net force acting on an object, what is it doing?

22. If there is a net force acting on an object, what is it doing?

23. What are the factors that determine the amount of frictional force between two materials?

24. How do you determine the components of an object's weight on an inclined plane?

25. If the distance between two masses triples, how does the gravitational force between them change?

26. For an object traveling in a circle at constant speed, what is the direction of the velocity, the acceleration, and the centripetal force?

27. At what angle should a projectile be fired for maximum range?

28. At what angle should a projectile be fired for maximum height?

29. What is the difference between momentum and impulse? How are they calculated? What are the units?

30. What is an elastic collision? What is conserved in an elastic collision?

31. What is an inelastic collision? What is conserved in an inelastic collision?

32. What is the work-energy theorem? How does it relate to conservation of energy?

33. Define critical angle. Is the critical angle in the more dense or less dense medium?

34. If the incident angle is less than the critical angle, what happens to the wave?

35. If the angle of incidence = the critical angle what happens to the wave?

36. If the angle of incidence is greater than the critical angle what happens?

37. As the angle of incidence increases, the critical angle_____.

38. What is the difference between transverse and longitudinal waves? Give examples for each.

39. What phenomenon proves light is transverse?

40. What phenomenon gave the wave theory preference over the particle theory?

41. Amplitude of a sound wave determines _____. Frequency determines _____.

42. Amplitude of a light wave determines _____. Frequency determines _____.

43. What are the differences between real and virtual images?

44. What are the image characteristics in a plane mirror?

45. What is the difference between a single and a double slit diffraction pattern?

46. Know how slit separation, distance to screen, wavelength, and distance to the first order maximum are related for double slit diffraction.

47. Explain the Doppler Effect for both sound and light.

48. What quantity stays constant when a wave passes from one medium to another?

49. What quantity stays constant when a wave travels through a non-dispersive medium?

50. Define: reflection, refraction, diffraction, interference, in and out of phase, amplitude, dispersion, and polarization.

51. Name seven kinds of EM waves in order of increasing wavelength.

52. What color of visible light has the longest wavelength?

53. What color of visible light has the highest frequency?

54. What change in an object must occur for it to become electrically charged?

55. Like charges _____, opposite charges _____.

56. To charge an object negatively by contact, the charging rod must be _____?

57. To charge an object negatively by induction, the charging rod must be _____?

58. If the distance between two charges doubles, how does the electrostatic force between them change?

59. What is an electric field? Know the electric field line diagrams for point charges and parallel plates.

60. What is Ohm's Law?

61. What quantity stays constant in a series circuit? In a parallel?

62. How is equivalent resistance calculated in a series circuit? In a parallel circuit?

63. How does equivalent resistance compare to any individual resistance in a series circuit? In a parallel circuit?

64. How are voltmeters and ammeters properly connected in a circuit?

65. What is the direction of the lines of force outside of a magnet? Inside? How can the relative strength of a magnet be represented by the magnetic field lines?

66. Two parallel wires carrying current in opposite directions _____ each other.

67. Explain the photoelectric effect. What is the threshold frequency?

68. Who did the gold foil experiment and what was his conclusion? What was Bohr noted for?

69. Write symbols for proton, neutron, electron, alpha, and beta particles.

70. Be able to use energy level diagrams for hydrogen and mercury on the reference tables.

71. What is a baryon?

72. What is a lepton?

73. What is the quark composition of a proton and a neutron?

74. Give five conservation laws.

75. Know the units for every variable listed on the reference tables.

PART I: MECHANICS AND ENERGY

UNIT 1
Mathematics and Vectors - A Toolkit for Physics

To fully understand the topics involved in a physics course, a certain amount of mathematical skill is required. These skills typically involve algebra, geometry, and trigonometry. Mathematics is considered a tool that is necessary in order to understand and communicate physics.

The Scale of Things...

Physics involves a study of the *macroscopic* and also the *microscopic* world. To effectively describe these sizes, measurements are expressed in the SI (*International System*) units. This system utilizes a modern form of the metric system. This system has **seven base units**, a multitude of **derived units**, and **metric prefixes** of multiples of ten to describe various quantities. The seven base units are as follows:

SI Base Units		
Quantity	**Unit**	**Symbol**
Length	meter	m
Mass	kilogram	kg
Temperature	kelvin	K
Time	second	s
Amount (of a substance)	mole	mol
Electric Current	ampere	A
Luminous Intensity	candela	cd

The SI units (also called MKS units) of length, mass, and time (base quantities) are *kilograms* (kg), *meters* (m) and *seconds* (s) respectively.

A derived unit contains a combination of any two or more base units. For example, the quantity of speed (or velocity) is measured using a derived unit. This unit is meters/second (m/s), which is a combination of the base units for length and time. Another example of a derived unit is the unit used to measure Force, a push or pull. The unit for force is the Newton (N). The Newton is equivalent to $kg \cdot \frac{m}{s^2}$, a combination of the base units of mass, length, and time squared. Many more derived units will be developed throughout the course.

1

Common Approximations

The following values are approximations of some common everyday items. These are not exact values and should not be used in direct calculations, but instead used to estimate values and/or determine whether a calculation made is reasonable.

Object	Value
Nickel (coin)	5 grams
1 Liter H_2O	1 kilogram
Car	1500 kg
Apple	1 Newton
Paperclip	1 gram
Fingers to nose distance	1 meter

Scientific Notation

Extremely large or extremely small numbers are often represented using scientific notation. To properly express a number in scientific notation, the number is to be at least one and less than ten, followed by the appropriate power of ten. For example:

$$15800 \text{ m} = 1.58 \times 10^4 \text{ m}$$
$$0.00000367 \text{ kg} = 3.67 \times 10^{-6} \text{ kg}$$

Order of Magnitude

An exponent represents the order of magnitude of a measurement when expressed in scientific notation. For example, 9140 N when expressed in scientific notation is 9.14×10^3 N. Since the exponent in scientific notation is a 3, the order of magnitude is 3 for this measurement.

Estimating the order of magnitude is especially useful when representing the approximate value of commonly found objects. For example, the approximate height of a high school student can easily be estimated as 10^0 m. Other orders of magnitude, such as 10^1 m or 10^{-1} m, would be far too inaccurate of a representation of this height. Even though a high school student is not 1 meter tall (10^0), the actual height might be 1.6×10^0 m, with an order of magnitude of 0.

Prefixes and Powers of 10

In order to represent numbers conveniently, metric prefixes are used to denote extremely large or small values. The following chart, from the **New York State Physics Reference Tables** shows common prefixes used in the SI system:

Prefixes for Powers of 10		
Prefix	Symbol	Notation
tera	T	10^{12}
giga	G	10^{9}
mega	M	10^{6}
kilo	k	10^{3}
deci	d	10^{-1}
centi	c	10^{-2}
milli	m	10^{-3}
micro	μ	10^{-6}
nano	n	10^{-9}
pico	p	10^{-12}

To use the above chart effectively, one must be able to convert from one prefix to another. This is done in the following method:

original value $($conversion factor(s)$)$ = desired value

For example, to convert from 258 km to meters:

$$258 \text{ km} \left(\frac{1 \times 10^3 \text{ m}}{1 \text{ km}} \right) = 2.58 \times 10^5 \text{ m}$$

When converting a more complex measurement, two or more conversion factors may be necessary.

For example, to convert from 94 cm to nm:

$$94 \text{ cm} \left(\frac{1 \times 10^{-2} \text{ m}}{1 \text{ cm}} \right) \left(\frac{1 \text{ nm}}{1 \times 10^{-9} \text{ m}} \right) = 9.4 \times 10^8 \text{ nm}$$

Measurement

When performing measurements of various quantities, careful attention must be paid so that the measurement is both *accurate* and *precise*.

Accuracy, or closeness to the actual value, is a result of proper measurement technique. The degree of accuracy can be expressed as a **percent difference**.

$$\% \text{ difference} = \left(\frac{\text{measured value - accepted value}}{\text{accepted value}} \right) \times 100$$

Precision, or the reproducibility of a measurement, refers to how often a measurement can be repeated for multiple measurements of the same object. For a single measurement, precision is determined by the device that is used to make the measurement. The measurement device is said to be precise if the measurement can be properly represented with a large number of **significant figures**. The more significant figures a measurement contains, the more precise it is, but not necessarily more accurate.

Measurement	Precision
45 s	1 second
1.2 m	1/10 of a meter
4.68 kg	1/100 of a kilogram

A common example used to understand accuracy and precision is that of a "bull's-eye," or a dartboard. If a single dart is thrown at a dartboard and it hits the bull's-eye, it is an accurate shot. If a number of darts are thrown, they can be categorized as accurate and/or precise as follows:

**Low Accuracy
High Precision**

**High Accuracy
Low Precision**

**High Accuracy
High Precision**

Measurement Technique

To be as accurate and precise as possible when making measurements, proper measuring technique must be utilized. When making measurements, the measurement must always be read *one decimal place beyond the smallest marking on the device.*

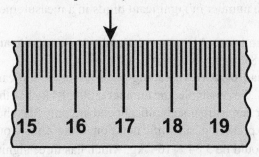

For example, when using a ruler or meter stick, as above, the smallest marking is typically the millimeter (mm). Therefore, when making measurements using this device the measurement is precise to the nearest 1/10 of a mm (0.0001 m). Using the ruler segment above, the measurement might then be read as: 16.72 cm. In this measurement, the first three digits (16.7) can be read directly from the device. The last digit (2) is estimated. It is not possible to record a measurement using this device to any more precision than this.

For other devices, such as digital multimeters and stopwatches, all of the numbers displayed are recorded and there is no estimation necessary. Using the stopwatch above, one would record the measurement as **16.12 s**.

A note about units...

At any point when a piece of data is to be recorded, the proper *unit* must be expressed with that measurement. It does not mean anything to record a measurement as 55.12. Recording the measurement as 55.12 s (seconds) identifies the measurement as having the dimension of time. Also, the units on numbers are treated like algebraic quantities when measurements are manipulated using arithmetic. For example, when multiplying 1.20 m by 2.30 m, the answer is 2.76 m². The units were multiplied together as well as the numbers.

Significant Digits

When data are expressed, this data must be represented with the correct number of **significant digits** (significant figures, or "sig figs"). It is important to be able to identify the number of significant digits of a measurement properly. The following rules are used to determine the number of significant digits in a measurement:

1. **Nonzero numbers:** <u>All</u> nonzero numbers are significant. For example, the number 1.458 cm has four significant digits.

2. **Initial, or leading zeros:** Leading zeros are <u>not</u> significant. These are simply placeholders. For example, the number 0.000584 kg has three significant digits. The first four zeros are not significant and are only placeholders. If the number were to be expressed in scientific notation, these zeros would disappear as well. The result would be 5.84×10^{-4} kg, which has three significant digits.

3. **Final, or trailing zeros:** Without a decimal, final zeros are <u>not</u> significant, such as in 520 m/s, which has two significant digits. With a decimal point, final zeros <u>are</u> significant, such as 520. m/s, which has three significant digits, or 146.1200 m, which has seven significant digits.

4. **Bound zeros:** Zeros between other significant digits <u>are</u> significant. For example, 808 N has three significant digits.

5. **Scientific Notation:** When a measurement is expressed in scientific notation, <u>all</u> digits are significant. For example, 5.91×10^{-7} m has three significant digits.

Arithmetic with Significant Digits

When performing mathematical operations, the final result cannot be any more significant than the least significant measurement. The following methods are used:

1. **Multiplication and Division:** When multiplying or dividing measurements, perform the operation, and then round your answer so that it has the same number of *significant digits* as the least significant measurement. For example:

$$5.40 \text{ cm} \times 2.571 \text{ cm} = 13.9 \text{ cm}^2$$

The result from a calculator would be 13.8834. Note that the result has been rounded to three significant digits, as the least number of significant digits is three, from the 5.40 cm measurement.

2. **Addition and Subtraction:** When adding or subtracting measurements, perform the operation, and then round your answer so that it has the same number of *decimal places* as the least precise measurement (the measurement with the least number of decimal places). For example:

$$5.1 \text{ kg} + 2.64 \text{ kg} = 7.7 \text{ kg}$$

Unrounded, the result from a calculator would be 7.74. Note that the result has been rounded to the nearest tenths place, as the least precise measurement of 5.1 kg is only precise to the tenths place.

Geometry and Trigonometry

There are many instances where the identities of geometry and trigonometry are utilized. Much of this information can be found in the Reference Tables as follows:

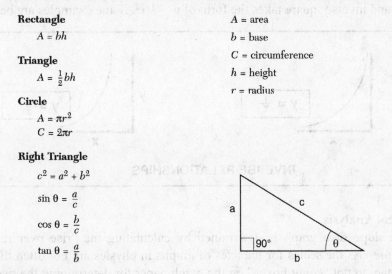

Rectangle

$A = bh$

Triangle

$A = \frac{1}{2}bh$

Circle

$A = \pi r^2$

$C = 2\pi r$

Right Triangle

$c^2 = a^2 + b^2$

$\sin \theta = \frac{a}{c}$

$\cos \theta = \frac{b}{c}$

$\tan \theta = \frac{a}{b}$

A = area

b = base

C = circumference

h = height

r = radius

Graphing Relationships

It is sometimes said that a picture is worth a thousand words. If that is the case, then a graph in physics must be worth a million. Graphically representing data can show what type of relationship exists among the data. It will also allow you to **extrapolate** or **interpolate** the data to be able to predict values that were not directly measured. **Extrapolation** of data involves predicting a value beyond the limits of those measured values, and **interpolation** involves predicting from within the data.

A **direct relationship** is one in which the quantity on the y-axis will increase as the quantity on the x-axis increases. There are two main types of direct relationships: the *direct linear* and also the *direct exponential* relationship. A direct linear graph is in the general form of $y = mx + b$, where m is the slope of the graph and b is the y-intercept. A direct exponential graph has the general form of $y = kx^n$, with k as a constant of proportionality, and n as the exponential power. Some examples are below:

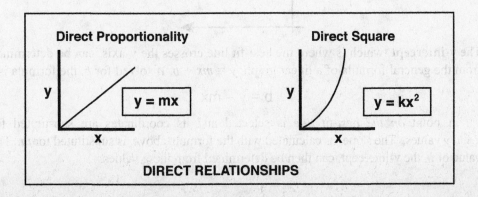

Direct Proportionality

$y = mx$

Direct Square

$y = kx^2$

DIRECT RELATIONSHIPS

An **inverse relationship** is one in which the quantity on the y-axis will decrease as the quantity on the x-axis increases. The two main types of inverse relationships are the *inverse* and the *inverse square*. In an inverse graph, the graph takes the general form of $y = 1/x$, and inverse square takes the form of $y = 1/x^2$. Some examples are below:

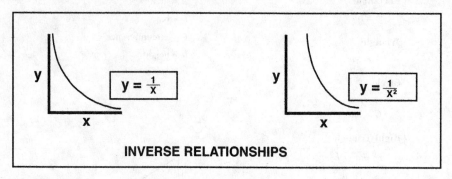

INVERSE RELATIONSHIPS

Graphical Analysis

The **slope** of a graph is determined by calculating the "rise over run" of the *best-fit line*. As the scales for the axes of graphs in physics are not often the same, it is important to not "count boxes" on the graph paper for determining the rise and run. Instead, the actual values of the points must be used. To do this, select two points that are *on the line*, which may or may not be data points, and calculate as follows:

$$\text{slope} = \frac{\Delta y}{\Delta x} = \frac{y_2 - y_1}{x_2 - x_1}$$

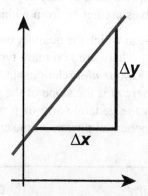

The **y-intercept**, which is where the best-fit line crosses the y-axis, can be determined from the general formula of a linear graph: $y = mx + b$. If solved for b, the formula is:

$$b = y - mx$$

A point *on the best-fit line* is selected and its coordinates are substituted for x and y values. The slope, as calculated with the formula above, is substituted for m. The value of b, the y-intercept, can then be determined from these values.

Calculating the **area under the curve** is another method of analyzing a graph in physics. With this method, the actual area between the best-fit line or curve and the x-axis is determined. This is done by breaking the graph into segments and using the formulas for area of a rectangle ($A = bh$) and the area of a triangle ($A = \frac{1}{2}bh$) to determine the total area. Again, because of the scales used on the axes, "counting boxes" on the graph paper will not give an accurate result for the area under the curve.

Sometimes, you may have to **predict values that are not data points.** To predict a value that is beyond the range of data collected is to **extrapolate** the data. To predict a value that is not a data point but is within the range of data collected is to **interpolate** the data. In either of these cases, it is best that you calculate the result using the equation of the line or curve of best fit rather than just reading the value off of the graph.

Scalar and Vector Quantities

All quantities in physics can be classified as either a **scalar quantity**, or a **vector quantity**. A scalar quantity is defined as a quantity that possesses *magnitude (size or amount) only*, whereas a vector quantity is a quantity that possesses both *magnitude and direction*.

To describe how far a person has traveled by saying "50 meters" is a scalar quantity of *distance*. However, to describe how far that person traveled by saying "50 meters west" is a vector quantity of *displacement*. Some common scalar and vector quantities in physics are listed below:

Scalar Quantities	Vector Quantities
distance	displacement
speed	velocity
time	acceleration
mass	force
energy	weight
work	momentum
power	impulse
charge	

Vector Representation

There are two ways to represent a vector quantity: either *graphically* or *algebraically*.

An *algebraic representation* uses numbers and words to describe the vector quantity, such as the following displacement vector:

$$20 \text{ m}, 30° \text{ West of North}$$

A *graphical representation* involves drawing the vector as an arrow pointing in the appropriate direction, and drawn to a scale. Vectors may be represented by using bold face type or with the symbol $\vec{v}$.

The length of the vector (arrow) is drawn proportional to the magnitude of the vector. The compass directions of north, south, east, and west are typically assumed to be consistent with that of a map.

Using the example above (20 m, 30° West of North), the vector would be drawn as follows:

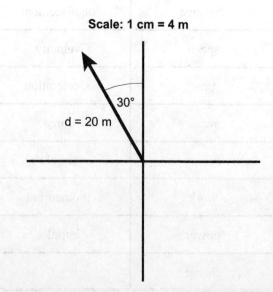

Scale: 1 cm = 4 m

30°

d = 20 m

Vector Addition

Vector addition is the process of combining two or more vector quantities. There are two main methods of vector addition: *graphical and algebraic.*

The *graphical method of vector addition* is sometimes called the **tip-to-tail method.** In this method the vectors being added together are arranged so that the tip (arrowhead end) of the first vector is touching the tail of the next, until all vectors are arranged. It is important to note that any vector can be "picked up and moved" anywhere on your paper as long as the two aspects that make it a vector, magnitude (length) and direction (the angle the vector points), are not changed. The **resultant (R)** is then drawn from the tail of the first vector to the tip of the last vector. See the example below:

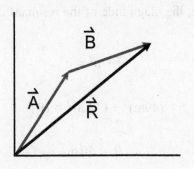

The **equilibrant (E)** is the vector that is equal in magnitude but opposite in direction to the resultant. It is a vector that creates **equilibrium** (a resultant of zero) with the other vectors. See the example below.

When adding three or more vectors together, the process is the same. Each vector is combined to the others so that they follow the tip-to-tail rule.

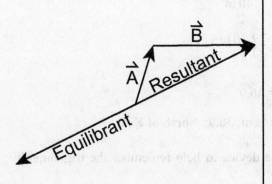

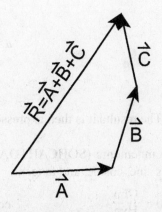

The *algebraic method* of vector addition is used when the two vectors being added together are at right angles to each other. In this situation, the two vectors and the resultant will form a right triangle. The Pythagorean Theorem ($a^2 + b^2 = c^2$) and trigonometry can be used to *calculate* the magnitude and direction of the resultant. See the example below:

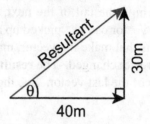

In the example above, the magnitude of the resultant can be determined by using the Pythagorean Theorem:

$$a^2 + b^2 = c^2$$

$$(40m)^2 + (30m)^2 = (R)^2$$

$$R = 50\,m$$

The direction of the resultant can be determined using one of the right triangle trigonometric functions (SOHCAHTOA). Since both the legs opposite and adjacent to the desired angle were given, the inverse of the tangent function can be used to solve for the angle.

$$\tan \theta = \frac{opp}{adj}$$

$$\tan \theta = \frac{30\,m}{40\,m}$$

$$\theta = \tan^{-1}\left(\frac{30\,m}{40\,m}\right)$$

$$\theta = 36.9°$$

The resultant is then expressed as **50m, 36.9° North of East**.

The mnemonic (SOHCAHTOA) is a device to help remember the trignometric functions sine, cosine, and tangent:

$$\sin \theta = \frac{Opp}{Hyp} \qquad \cos \theta = \frac{Adj}{Hyp} \qquad \tan \theta = \frac{Opp}{Adj}$$

The angle between vectors is illustrated when the vectors are **tail to tail**. For example, when the angle between two vectors is 0°, the two vectors are pointing in the same direction. This will produce a resultant with the largest magnitude. Mathematically, the magnitude of the resultant is found by adding the magnitudes of each vector.

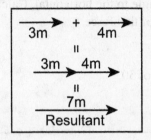

When the angle between two vectors is 180°, the two vectors are in the opposite direction, and will produce a resultant with the smallest magnitude. The magnitude of the resultant can be found by subtracting the magnitude of one vector from the other.

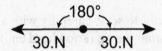

If the directions of vectors can be changed, the range of resultants can be determined by finding the maximum resultant (when arranged at 0°), and the minimum resultant (when arranged at 180°). When the vectors are arranged at any other angle, the resultant will be between these values. For example, the range of resultants for a 15.0 N force and a 20.0 N force, arranged at any angle, will be from 5.0 N to 35.0 N. Regardless of the angle between these two vectors, the resultant can not be smaller than 5.0 N and can not be larger than 35.0 N.

Vector Components

The components of a vector are any two (or more) vectors that, when added together, form the original vector. One can think of finding the components of a vector as a vector addition problem in reverse. Imagine that the resultant (the vector for which you want to find the components) is provided and you ask yourself what vectors added together will make that resultant.

Although there is <u>an infinite number of components of a vector</u>, the components most useful are the <u>perpendicular components</u>. The perpendicular components are the vertical (y-direction) and horizontal (x-direction) vectors that combine to form the original vector. The perpendicular components can be found with either the graphical or algebraic methods. In the algebraic method, the following equations are used:

$$A_y = A \sin \theta$$
$$A_x = A \cos \theta$$

Solved Example Problem

- A force of 40. N is required to pull a wagon across the road. The force is applied at a 50.° angle to the horizontal. Calculate the magnitude of the x and y components of this force.

$A_x = A \cos \theta$ $A_y = A \sin \theta$

$A_x = 40.\text{N} \cos 50.°$ $A_y = 40.\text{N} \sin 50$

$A_x = 26 \text{ N}$ $A_y = 31 \text{ N}$

To determine the perpendicular components of a vector graphically, the following steps should be followed:

1. Draw the original vector to scale carefully with a ruler and protractor.
2. Draw a vertical line from the tip (arrowhead end) of the original vector.
3. Draw a horizontal line from the tail of the original vector.
4. Erase any excess length of the vertical and horizontal lines, and add arrowheads onto the components.
5. Measure the lengths of these components considering the scale used for the original vector drawing.

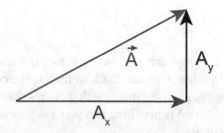

Unit 1
Practice Questions

1. The weight of a typical high school physics student is closest to

   (1) 1500 N   (2) 600 N   (3) 120 N   (4) 60 N

2. The speedometer in a car does *not* measure the car's velocity because velocity is a

   (1) vector quantity and has a direction associated with it.
   (2) vector quantity and does not have a direction associated with it.
   (3) scalar quantity and has a direction associated with it.
   (4) scalar quantity and does not have direction associated with it.

3. An airplane flies with a velocity of 750. kilometers per hour, 30.0° south of east. What is the magnitude of the eastward component of the plane's velocity?

   (1) 866 km/h   (2) 650. km/h   (3) 433 km/h   (4) 375 km/h

4. The mass of a paper clip is approximately

   (1) 1×10^6 kg   (2) 1×10^3 kg   (3) 1×10^{-3} kg   (4) 1×10^{-6} kg

5. The diagram below represents two concurrent forces.

Which vector represents the force that will produce equilibrium with these two forces?

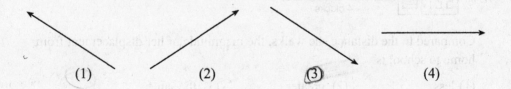

   (1)   (2)   (3)   (4)

6. As the angle between two concurrent forces decreases, the magnitude of the force required to produce equilibrium

    (1) decreases   (2) increases   (3) remains the same

7. A child walks 5.0 meters north, then 4.0 meters east, and finally 2.0 meters south. What is the magnitude of the resultant displacement of the child after the entire walk?

    (1) 1.0 m   (2) 5.0 m   (3) 3.0 m   (4) 11.0 m

8. The diagram below represents a force vector, A, and a resultant vector, R.

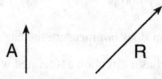

Which force vector B below could be added to force vector A to produce resultant vector R?

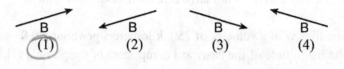

    (1)   (2)   (3)   (4)

---

9. A student on her way to school walks four blocks east, three blocks north, and another four blocks east, as shown in the diagram.

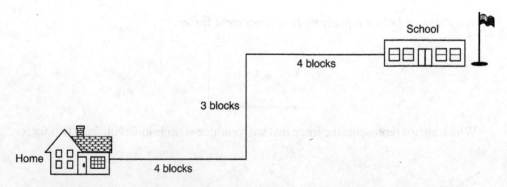

Compared to the distance she walks, the magnitude of her displacement from home to school is

    (1) less   (2) greater   (3) the same

---

10. Two 30.-newton forces act concurrently on an object. In which diagram would the forces produce a resultant with a magnitude of 30. newtons?

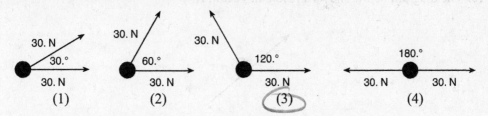

(1)   (2)   (3)   (4)

11. A 5.0-newton force could have perpendicular components of

(1) 1.0 N and 4.0 N   (3) 3.0 N and 4.0 N
(2) 2.0 N and 3.0 N   (4) 5.0 N and 5.0 N

12. The vector diagram below represents two forces, F_1 and F_2, simultaneously acting on an object.

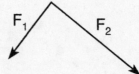

Which vector best represents the resultant of the two forces?

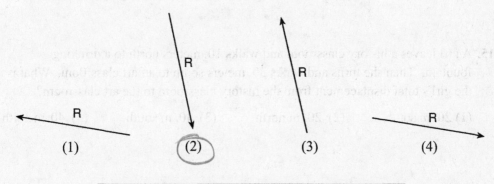

(1)   (2)   (3)   (4)

---

13. An egg is dropped from a third-story window. The distance the egg falls from the window to the ground is closest to

(1) 10^0 m   (2) 10^1 m   (3) 10^2 m   (4) 10^3 m

14. The diagram below shows a resultant vector, R.

Which diagram best represents a pair of component vectors, A and B, that would combine to form resultant vector R?

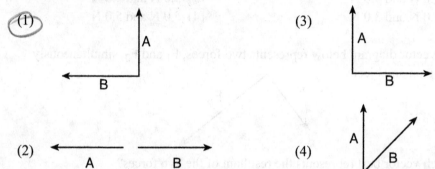

15. A girl leaves a history classroom and walks 10. meters north to a drinking fountain. Then she turns and walks 30. meters south to an art classroom. What is the girl's total displacement from the history classroom to the art classroom?

(1) 20. m south   (2) 20. m north   (3) 40. m south   (4) 40. m north

16. A vector makes an angle, θ , with the horizontal. The horizontal and vertical components of the vector will be equal in magnitude if angle θ is

(1) 30°   (2) 45°   (3) 60°   (4) 90°

17. Which pair of forces acting concurrently on an object will produce the resultant of greatest magnitude?

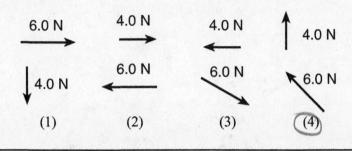

18. Which term represents a scalar quantity?

(1) distance   (2) displacement   (3) force   (4) weight

19. Which is a vector quantity?

(1) distance   (2) speed   (3) power   (4) force

20. A force vector was resolved into two perpendicular components, F_1 and F_2, as shown in the diagram below.

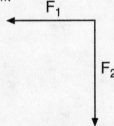

Which vector best represents the orginal force?

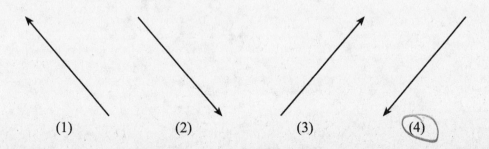

NOTES:

UNIT 2
Kinematics: Constant and Changing Velocities

Kinematics is the description of how objects move without regard for the cause of motion.

One-dimensional Motion

For an object to be in motion there must be a change in **position** of the object with respect to **time**. If an object is moving in a straight line, it is said to be in **one-dimensional motion.**

Speed and Velocity

Speed is a scalar quantity that represents how fast an object is traveling. It is determined by dividing the distance traveled by the time it takes to travel that distance.

$$\text{speed} = \frac{\text{total distance}}{\text{total time}}$$

$$\overline{v} = \frac{d}{t}$$

Note that the line over the v in the equation above denotes the average speed. The typical unit of distance in the SI System is the **meter (m)**, and the typical unit of time is the **second (s)**. Therefore, the typical unit of measure for speed is **meters per second (m/s)**.

Velocity is a vector quantity that represents how fast an object is moving and in which direction it is traveling. The **magnitude of the velocity** is the speed of an object. Velocity is calculated in a similar fashion to speed; however, the displacement is divided by the total time.

$$\text{velocity} = \frac{\text{displacement}}{\text{total time}}$$

$$\overline{v} = \frac{d}{t}$$

As with speed, the unit of velocity is meters per second (m/s). Note in the equation above that the **boldface** letters denote **vector quantities.**

Solved Example Problems

1. An object travels for 8.00 seconds with an average speed of 160. meters per second. Calculate the distance traveled by this object.

given information

$t = 8.00s$

$v = 160.$ m/s

$d = ?$

$\bar{v} = \dfrac{d}{t}$

$d = \bar{v}t$

$d = 160.$ m/s $\cdot$ 8.00s

$d = 1280$ m

2. The average speed of a runner in a 400. meter race is 8.0 meters per second. How long did it take the runner to complete the race?

given information

$d = 400.$ m

$v = 8.0$ m/s

$t = ?$

$\bar{v} = \dfrac{d}{t}$

$t = \dfrac{d}{\bar{v}} = \dfrac{400.\,m}{8.0\,m/s}$

$t = 50.$ s

Accelerated Motion

Acceleration is the rate of change of velocity. Acceleration is a vector quantity, having both magnitude and direction. Acceleration can be calculated by dividing the change in velocity by the time elapsed.

$$\text{acceleration} = \frac{\text{change in velocity}}{\text{time}}$$

$$a = \frac{\Delta v}{t}$$

In the equation above the letter *delta* (Δ) refers to a change in a quantity. In this case, it is referring to the change in velocity. The term Δv can be expanded to $v_f - v_i$. The unit of velocity is meters per second (m/s) and the unit of time is seconds (s); therefore, the unit of acceleration is meters per second per second (m/s/s), or **meters per second squared (m/s^2)**.

Solved Example Problem

• An object moving in a straight line at an initial speed of 6.0 m/s accelerates uniformly for 5.0 s to a final speed of 36 m/s. Find the object's acceleration during this time interval.

given information

$v_i = 6.0$ m/s

$v_f = 36$ m/s

$t = 5.0$ s

$a = ?$

$a = \dfrac{\Delta v}{t}$

$a = \dfrac{36\frac{m}{s} - 6.0\frac{m}{s}}{5.0\,s}$

$a = 6.0$ m/s^2

Kinematics Equations

In addition to the basic equations for velocity and acceleration, additional equations can be derived that can be used to solve for various properties of an object's motion. These **kinematics equations** are as follows:

$$d = v_i t + \frac{1}{2} at^2$$

$$v_f^2 = v_i^2 + 2ad$$

$$v_f = v_i + at$$

where:
d = distance or displacement
v_i = initial velocity
v_f = final velocity
t = time
a = acceleration

Free-Fall

When an object is undergoing free-fall, the object is falling toward the earth without any forces acting on it except for gravity. This is best observed in a vacuum, where there are no effects from air resistance. *In a vacuum*, all objects, regardless of mass, fall toward the earth with the same acceleration: the **acceleration due to gravity.**

$$a_{\text{free-fall}} = g = 9.81 \text{ m/s}^2$$

The above value means that an object will increase its speed by 9.81 m/s for each second of time that object is falling. After one second of fall, it will be traveling 9.81 m/s, after two seconds, 19.62 m/s, and so on.

If it is known that the object is undergoing free-fall, then the value of g (9.81 m/s^2) can be listed as one of the given values in a kinematics problem. All of the equations of kinematics can be used to solve free-fall problems.

In everyday situations, it is difficult to prove that all objects accelerate at 9.81 m/s^2 because of the presence of air resistance. For example, if you were to drop a penny and a feather toward the ground, the penny will most certainly hit the ground first. This is because the air resistance on the feather will have a significant effect on the acceleration, and not as much effect on the penny. However, if the same demonstration were to be done in a device with the air removed (in a vacuum chamber), then the two objects would accelerate toward the earth's surface at the same rate: 9.81 m/s^2!

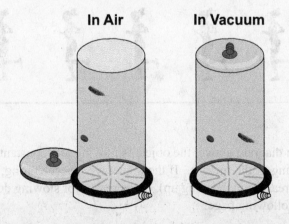

In Air   **In Vacuum**

An experiment similar to this was also performed on the moon. In 1971, Commander David Scott of the Apollo 15 mission released a geologic hammer and a falcon feather at the same time to fall toward the lunar surface. As the moon has no atmosphere, there was no air resistance and, as predicted, they both fell to the moon's surface and landed at the same time. The moon has a different value for its acceleration due to gravity, approximately 1.62 m/s^2, so the objects did not fall as fast as if it were done on the earth, but they did fall together as predicted!

A Note about Direction

Since both velocity and acceleration are vector quantities, designating the direction of these quantities is important as well. When working with kinematics equations, the direction of the quantities is represented with the proper sign. Usually, *up* is designated a positive direction, and *down* is designated as negative. With this convention, the value of g will then be -9.81 m/s^2. However, if down is positive, then up must be negative. Likewise, *to the right* is typically positive, and *to the left* is typically negative. These sign conventions must be done consistently, especially with problems involving the acceleration due to gravity.

Motion Diagrams and Particle Diagrams

A **motion diagram** is a series of images of an object layered, or superimposed, to depict the motions of the object. One can think of a photo taken with a strobe light as an example. It is important that the time elapsed between images be constant to accurately represent the motion. In addition to the overall movements of an object, a motion diagram may also depict the smaller motions. For example, the spacing of a bird in flight will illustrate whether the bird is flying at constant speed. The movements of the wings may also be illustrated to provide even more detail of the motion of the bird. If the motion diagram is to scale, and the time elapsed between images is known, numerical calculations may also be made to determine the speed and acceleration of the object. An example of a motion diagram is below:

In the motion diagram above, the object is moving at a constant speed as depicted by the equal spacing of the images. If the object were accelerating, the spacing of the objects would increase (for speeding up), or decrease (for slowing down) as seen in the examples on the following page:

Less detailed than a motion diagram is a **particle diagram**. In a particle diagram, essentially all of the fine detail of the object and surroundings are removed and a dot is used to represent the object. The dot is always drawn at the same location on the object for each position. It is typically drawn at the **center of mass** of the object. Like a motion diagram, the spacing of the dots will determine the characteristics of the motion of the object.

· · · · · · · · ·

In the particle diagram above, the object is moving with some constant velocity, as noted by the equal spacing of the dots. However, without additional information about the object, it is not known in which direction the object is traveling.

·· · · · · · ·

The object in the diagram above is accelerating. There are two possible scenarios here. Either the object could be traveling to the right and speeding up, or this object could be traveling to the left and slowing down. Again, without additional information, each scenario is equally plausible.

In these situations, the sign of the velocity and acceleration are essential in order to properly describe the motion. Knowing the sign of the velocity, for example, will designate the direction of travel. Typically, the positive direction is to the right and negative is to the left. In the particle diagram above, the object may have had a positive velocity and a positive acceleration. This would indicate that it was moving to the right and speeding up. If, however, the object had a negative velocity and positive acceleration, it would have instead been traveling to the left and slowing down.

•   •   •   •   •   • •

In the particle diagram above, the object may have had a positive velocity and a negative acceleration. In this case the object would have been traveling to the right and slowing down. However, it may have had a negative velocity and a negative acceleration instead. This would mean that the object was traveling to the left and speeding up. Like in the other examples, the actual motion is not known without additional information. Sometimes an arrow is added to the diagram to indicate the direction of travel, which would help clarify the situation.

Kinematics Problem Solving

With all of these equations, and all of the quantities represented by them, it is important to follow a problem solving procedure to select the proper equation to use and properly substitute the values into it. Most of the kinematics problems that must be solved are in the form of word problems. Properly solving a word problem in physics is essential for success. The following steps are useful to properly solve kinematics problems:

1. Read the problem carefully. Make sure you understand all the terms used.
2. Wherever possible produce a sketch of the situation described. Label all the given quantities on your sketch.
3. Carefully make a list of all the quantities given in the problem using the letter symbol and setting it equal to the value given.
   ex.:
   $$v_f = 4 \text{ m/s}$$
   $$t = 5 \text{ s}$$
4. Add to the list of "givens" any other quantities that may be determined from the description in the problem but which are not specifically stated.
   ex.:
   $$a = g = 9.81 \text{ m/s}^2 \text{ (a falling object)}$$
   $$v_i = 0 \text{ (started from rest)}$$
5. Finally, indicate on this list the symbol for the variable(s) to be determined with a question mark.
   ex.:
   $$d = ?$$
6. Inspect the list and find the equation on your equation sheets or reference tables that fits the variables. Write out this equation.
7. Plug in the data and solve the equation for the variable to be determined. Be sure to include the unit for each quantity in the equation. NOTE: If this process is not working, a sub-calculation may be needed to arrive at values to use in the final calculation.
8. Inspect your final answer. Do the units make sense? Is the magnitude of the answer reasonable for the situation described?

Solved Example Problem

If you dropped a stone from a cliff and it took 5.00 seconds to hit bottom, how fast was it going when it hit?

Variables:

$v_i = 0$

$t = 5.00$ s

$g = -9.81$ m/s^2

$v_f = ?$

Equation: $v_f = v_i + at$

Substitution with units:

$v_f = 0 + (-9.81$ m/s$^2)(5.00$ s$)$

Work & answer with units:

$v_f = -49.1$ m/s

(the negative denotes the direction of the final velocity is downward)

PRACTICE PROBLEMS FOR KINEMATICS PROBLEM SOLVING

1. If you ran 1500 meters in 10. minutes, what is your average speed in m/s?

   Variables: Equation:

                                                 Substitution with units:

                                                 Work & answer with units:

2. A car accelerates from 20.0 m/s to 60.0 m/s in 4.00 seconds. What was the average acceleration of the car and how far did it travel during the 4.00 seconds?

   Variables: Equation:

                                                 Substitution with units:

                                                 Work & answer with units:

3. If a car accelerated past you with its speed changing from 15.0 m/s to 25.0 m/s in 5.0s,

   a) What was its average acceleration?

   b) How far would it travel during this time?

Graphical Analysis of Motion

In addition to solving problems using the equations of kinematics, creating and interpreting graphs of motion can be helpful and informative. There are three main graphs of motion:

   1. displacement vs. time (d vs. t)
   2. velocity vs. time (v vs. t)
   3. acceleration vs. time (a vs. t)

It is important to note that regardless of whether it is the dependent or independent variable, **time is always on the x-axis** of these motion graphs. Some examples of motion graphs are below:

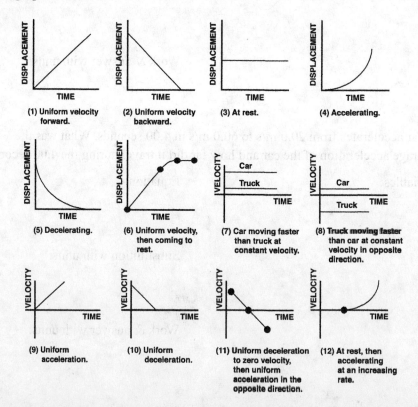

CONSTANT AND CHANGING VELOCITIES

When given one of the graphs of motion, one is often asked to determine another quantity from it. For example, if given a velocity vs. time graph, the acceleration can be determined. This is done by calculating the **slope** of the graph. Likewise, the average velocity can be determined by calculating the **slope** of a linear displacement vs. time graph. If a displacement vs. time graph is not linear, the **slope of a tangent line**, tangent to the curve at the desired time, will determine the *instantaneous velocity*.

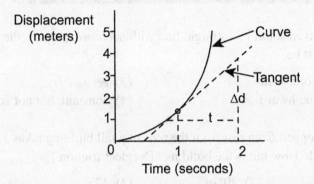

The total distance traveled can be determined by calculating the **area under the curve** of a velocity vs. time graph. To calculate area under the curve, one simply calculates the geometric area of the space between the line (or curve) and the x-axis of the graph. This is typically done by breaking up the area into either triangles or rectangles and using the equations for area of a rectangle (Area=bh) or a triangle (Area $= ^1/_2$ bh). It is important to use the proper scales on the graph axes when calculating area under the curve and not to count boxes on the graph paper.

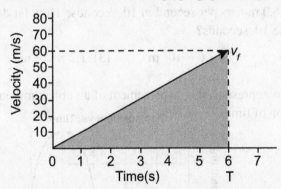

The chart below is a useful way to remember which method to use for graphical analysis:

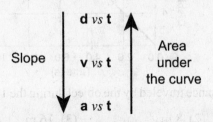

29

Unit 2
Practice Questions

1. As a car is driven south in a straight line with *decreasing* speed, the acceleration of the car must be

   (1) directed northward   (3) zero
   (2) directed southward   (4) constant, but not zero

2. A baseball dropped from rest from the roof of a tall building takes 3.1 seconds to hit the ground. How tall is the building? [Neglect friction.]

   (1) 15 m   (2) 30. m   (3) 47 m   (4) 94 m

3. A rock falls from rest a vertical distance of 0.72 meter to the surface of a planet in 0.63 second. The magnitude of the acceleration due to gravity on the planet is

   (1) 1.1 m/s²   (2) 2.3 m/s²   (3) 3.6 m/s²   (4) 9.8 m/s²

4. The speed of an object undergoing constant acceleration increases from 8.0 meters per second to 16.0 meters per second in 10. seconds. How far does the object travel during the 10. seconds?

   (1) 3.6×10^2 m   (2) 1.6×10^2 m   (3) 1.2×10^2 m   (4) 8.0×10^1 m

5. The graph below represents the displacement of an object moving in a straight line as a function of time.

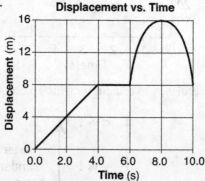

Displacement vs. Time

What was the total distance traveled by the object during the 10.0-second time interval?

   (1) 0 m   (2) 8 m   (3) 16 m   (4) 24 m

6. Which graph best represents the relationship between the velocity of an object thrown straight upward from Earth's surface and the time that elapses while it is in the air? [Neglect friction.]

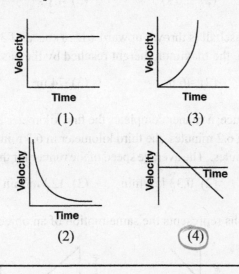

7. A car increases its speed from 9.6 meters per second to 11.2 meters per second in 4.0 seconds. The average acceleration of the car during this 4.0-second interval is

(1) 0.40 m/s²   (2) 2.4 m/s²   (3) 2.8 m/s²   (4) 5.2 m/s²

8. A cart travels with a constant nonzero acceleration along a straight line. Which graph best represents the relationship between the distance the cart travels and time of travel?

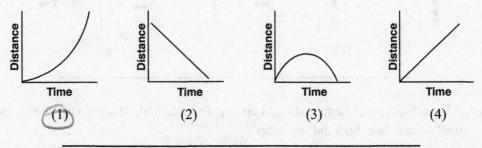

9. A rocket initially at rest on the ground lifts off vertically with a constant acceleration of 2.0×10^1 meters per second². How long will it take the rocket to reach an altitude of 9.0×10^3 meters?

(1) 3.0×10^1 s   (2) 4.3×10^1 s   (3) 4.5×10^2 s   (4) 9.0×10^2 s

10. A 1.0-kilogram ball is dropped from the roof of a building 40. meters tall. What is the approximate time of fall? [Neglect air resistance.]

(1) 2.9 s   (2) 2.0 s   (3) 4.1 s   (4) 8.2 s

11. A 0.25-kilogram baseball is thrown upward with a speed of 30. meters per second. Neglecting friction, the maximum height reached by the baseball is approximately

(1) 15 m   (2) 46 m   (3) 74 m   (4) 92 m

12. In a 4.0-kilometer race, a runner completes the first kilometer in 5.9 minutes, the second kilometer in 6.2 minutes, the third kilometer in 6.3 minutes, and the final kilometer in 6.0 minutes. The average speed of the runner for the race is approximately

(1) 0.16 km/min   (2) 0.33 km/min   (3) 12 km/min   (4) 24 km/min

13. Which pair of graphs represents the same motion of an object?

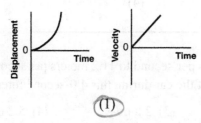

(1)

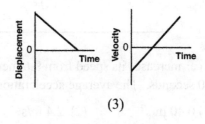

(3)

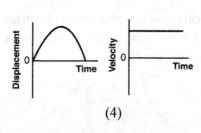

(2)

(4)

---

14. A basketball player jumped straight up to grab a rebound. If she was in the air for 0.80 second, how high did she jump?

(1) 0.50 m   (2) 0.78 m   (3) 1.2 m   (4) 3.1 m

15. One car travels 40. meters due east in 5.0 seconds, and a second car travels 64 meters due west in 8.0 seconds. During their periods of travel, the cars definitely had the same

(1) average velocity   (3) change in momentum
(2) total displacement   (4) average speed

16. A ball thrown vertically upward reaches a maximum height of 30. meters above the surface of Earth. At its maximum height, the speed of the ball is

(1) 0.0 m/s   (2) 3.1 m/s   (3) 9.8 m/s   (4) 24 m/s

17. How far will a brick starting from rest fall freely in 3.0 seconds?

(1) 15 m   (2) 29 m   (3) 44 m   (4) 88 m

18. A roller coaster, traveling with an initial speed of 15 meters per second, decelerates uniformly at -7.0 meters per second² to a full stop. Approximately how far does the roller coaster travel during its deceleration?

(1) 1.0 m   (2) 2.0 m   (3) 16 m   (4) 32 m

19. The graph below shows the velocity of a race car moving along a straight line as a function of time.

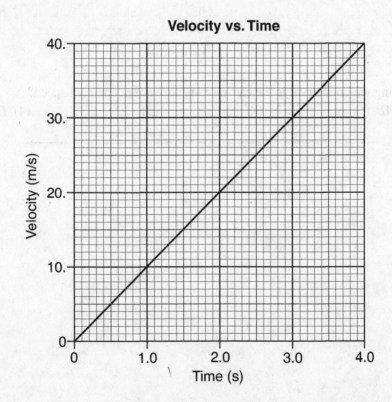

Velocity vs. Time

What is the magnitude of the displacement of the car from t = 2.0 seconds to t = 4.0 seconds?

(1) 20. m   (2) 40. m   (3) 60. m   (4) 80. m

20. The displacement-time graph below represents the motion of a cart initially moving forward along a straight line.

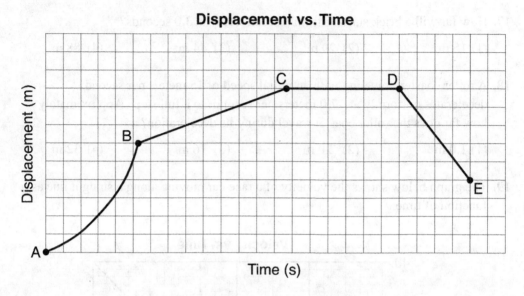

Displacement vs. Time

During which interval is the cart moving forward at constant speed?

(1) *AB*   (2) *BC*   (3) *CD*   (4) *DE*

_____

UNIT 3
Dynamics: Forces and Newton's Laws

In the previous unit, the motion of an object was described without taking into consideration the agent(s) causing these motions or changes in motion. In this unit, the concepts of force will be reviewed, which describes the causes of these motions or changes in motion.

A **force** is any push or pull on an object. Force is a **vector quantity**, having both magnitude and direction. The relationship between force and the motions of an object is called **dynamics**.

Contact and Field Force

A **contact force** is any force that is the result of two or more objects physically making contact with each other. Some examples of contact forces include: pushing a cart across the table, a chair holding a person up, pushing a swing, and pulling a sled across the snow.

A **field force**, or non-contact force, is any force acting on an object at various positions without physical contact. If you were to bring two magnets near each other, they would either attract or repel, depending on the orientation of the poles. This attractive or repulsive force is a result of the magnetic field created by the magnets. The magnets are not actually touching each other, but their magnetic fields result in the force between them. Some other examples of field forces include: forces from electric fields, gravitational fields, the strong nuclear force, and the weak force.

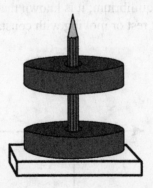

Diagramming Forces

A **free-body diagram, (FBD),** is a diagram of an object to be analyzed with all of the forces acting upon that object labeled. As forces are vectors, they must be drawn to relative scale, and pointing in the appropriate direction. Often, the FBD removes much of the detail of the object and replaces it with a single dot, again at the center of mass of the object. Also, the force vectors are drawn to originate from the object, with all of the forces pointing away from it.

By analyzing a free-body diagram, the **net force**, or vector sum of the forces, can be determined. If the net force is known, the motion, if any, of the object can be determined. If the net force on an object is zero, the object is in **equilibrium**. An object in equilibrium has an acceleration of zero. That means that the object may be at rest or moving in a straight line with constant speed. If the net force is not zero, the object has an **unbalanced force** acting on it. This unbalanced force will change the magnitude and/or direction of the objects velocity, therefore that object must be accelerating.

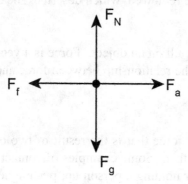

In the free-body diagram above, the following forces are depicted: the Normal Force (F_N) which is the support force between a surface (such as a table or the floor) and the object; the Applied Force (F_a) which may be an external push or pull on the object (like someone pushing a box); the Force of Gravity (F_g), or weight, from the earth; and the Force of Friction (F_f), which is the result of two surfaces rubbing against each other. In the diagram above, the net force in the horizontal (x) direction is zero because the two vectors are the same length and in the opposite direction. Likewise, the net force in the vertical (y) direction is also zero. Therefore, the overall net force on this object is zero. Since this object is in equilibrium, it is known that it has an acceleration of zero. Therefore, it must be either at rest or moving with constant speed in a straight line.

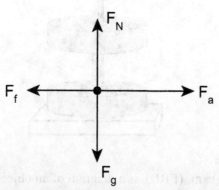

In the FBD above, depicting the same four forces, the net force is no longer zero. The applied force vector is now longer than the friction force vector, so these forces do not cancel each other out. As a result, the net force in the horizontal direction is not zero. In the vertical direction, the net force remains zero. Since there is now a net force on the object, it will now have an acceleration.

Newton's Laws of Motion

Sir Issac Newton

Sir Isaac Newton was an English physicist, mathematician, and astronomer most noted for being the "father of classical mechanics" by describing his Three Laws of Motion, the Law of Universal Gravitation, and much more. The SI unit of force (the Newton, N) is named after him.

Born: January 4, 1643; England
Died: March 31, 1727; England

College Attended: Trinity College, Cambridge

1. Newton's First Law: Often called the Law of **Inertia**, this law describes the tendency of an object to resist any changes to its motion or state of rest. A description of the law of inertia is as follows: An object at rest tends to stay at rest and an object in motion tends to stay in motion at constant speed in a straight line, unless there is an unbalanced force acting upon it. Inertia is directly proportional to the object's mass. It is more difficult to change the motion of an object with greater mass. Therefore, it has greater inertia. It is much easier to catch a tennis ball thrown toward you than it is to catch a bowling ball. The greater mass of the bowling ball requires more force to change its motion (stop it) than is necessary for the tennis ball.

Using the law of inertia, the coin in the image above can be dropped into the beaker without touching it. If one were to give the cardboard under the coin a quick pull out of the way, the coin, an object at rest, will fall right into the beaker! Because of inertia, the coin will stay at rest when the card is quickly removed. Since the card is now gone, there is nothing supporting the coin and it will fall into the beaker. As long as it is done quickly, the friction between the coin and the card will not be enough to push the coin out of the way.

Another common demonstration of the law of inertia is the tablecloth demonstration. In this example, the tablecloth is pulled out from under a set of dishes without the dishes crashing to the floor. This demonstration requires specific conditions to work properly but, if done correctly, the dishes will stay in place!

2. Newton's Second Law: Sometimes referred to as the Law of Force and Acceleration, this law states if there is an unbalanced (net) force acting on an object, that object will accelerate in the direction of that net force. The acceleration is directly proportional to the net force, and inversely proportional to the mass of the object. It is expressed in equation form as follows:

$$a = \frac{F_{net}}{m} \qquad \text{or} \qquad F_{net} = ma$$

In the equations above:
- a = acceleration - measured in m/s^2
- F_{net} = net (unbalanced) force - measured in N (Newtons)
- m = mass - measured in kg

Note that taken directly from the equation, the units of F_{net} would be: $\frac{kg \cdot m}{s^2}$.

Therefore, it can be stated that $1N = 1\frac{kg \cdot m}{s^2}$

An example to assist in understanding the relationships between these variables is that of a broken-down car. Imagine if a car full of people were to break down in the road. As there is no other traffic on this deserted road, it is safe to get out of the car to push it to the shoulder. According to $a = \frac{Fnet}{m}$, in order to obtain the largest acceleration to move the car, one must have the largest force, and the smallest mass. The best way to do this is to ask everyone in the car to get out and help push. This has two effects: it reduces the mass of the vehicle and its contents, and it maximizes the net force applied to the car. This will result in the greatest acceleration of the car to get it to the side of the road most efficiently. Solving for the different variables correctly in Newton's Second Law is essential for success in this topic. One way to remember how to rearrange the variables in the equation is to use the force triangle pictured below.

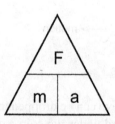

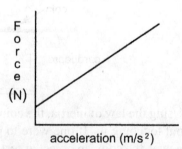

acceleration (m/s^2)

In using the force triangle, cover the unknown variable with your finger or thumb. The remaining letters and their arrangement will show whether you need to multiply or divide. For example, to solve for mass (m), cover the *m* and what remains is F/a. This tells you that in order to solve for *m*, you must divide the net force by the acceleration.

Force and acceleration can also be plotted on a graph, as in the example above. When plotted with Force on the y-axis and acceleration on the x-axis, the slope of this graph equals the mass of the object being accelerated. Since mass is constant, the graph of **F vs. a** is a straight line.

3. Newton's Third Law: Often called the Law of Action-Reaction, this law states that for every action (force) there is an equal (in magnitude) and opposite (in direction) reaction (force). It is important to note that these action-reaction pairs of forces are acting on *different* objects and therefore will not appear on the same free-body diagram. For example, imagine while driving in a car a bug suddenly hits the windshield. One could consider the action force the windshield hitting the bug. The reaction force then is the bug hitting the windshield.

These forces are equal in magnitude and opposite in direction. There is significant damage caused to the bug, but essentially no effect to the windshield. This is because of the significant difference in masses. Referring to Newton's Second Law, the mass is inversely proportional to the acceleration, given a constant force. Since the mass of the bug is so small, the acceleration of the bug is very, very large. Since the mass of the car is so large, there is an immeasurably small acceleration of the car, and it is hardly noticed except for a small sound produced on impact.

A rocket on liftoff is another example demonstrating Newton's Third Law. In this situation, the action force is the rocket engine's push of the hot combustion gases from the bottom of the rocket in the downward direction. The reaction force is the upward push of the hot gases on the rocket. Because of this unbalanced force on the rocket in the upward direction, the rocket accelerated up and is lifted off.

Weight

Weight is the force on an object from gravity, most commonly the force exerted on an object from the gravitational pull of the earth. The weight of an object on earth can be calculated using a modified version of Newton's Second Law:

$$F_g = mg$$

In the equation above, F_g represents the weight, measured in Newtons; m is the mass, measured in kilograms; and g is the acceleration due to gravity, typically 9.81 m/s^2 for earth. Weight is a vector quantity with both magnitude and direction. Although the direction is not often explicitly stated, the direction is straight down toward the center of the earth. On other planets, the weight of an object will be different, but its mass will always remain constant. This difference in weight is due to the fact that g is different on different planets.

Newton's Law of Universal Gravitation

All objects with mass exhibit a mutual attraction toward one another. This attraction is called **gravitational attraction**. Isaac Newton first published this in his work *Philosophiae Naturalis Principia Mathematica* (the "Principia"), in 1687. The magnitude of this gravitational force was found to be directly proportional to the product of the masses of each object and inversely proportional to the square of the distances between their centers. This is an example of an *inverse-square law*.

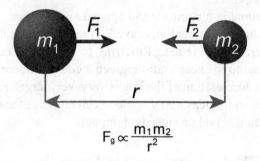

$$F_g \propto \frac{m_1 m_2}{r^2}$$

It was determined that the gravitational force was not exactly equal to the products of the masses divided by the square of the distance between their centers. There was a constant of proportionality that was missing. In 1798 Henry Cavendish performed an experiment in which the constant of proportionality was determined. In this experiment, Cavendish suspended masses using a torsion balance and measured the gravitational attraction between these masses.

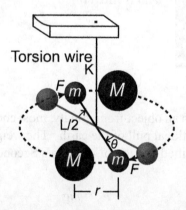

With the determination of the Universal Gravitational Constant (G), Newton's Law of Universal Gravitation becomes:

$$F_g = G \frac{m_1 m_2}{r^2},$$

where $G = 6.67 \times 10^{-11} \ Nm^2/kg^2$

Since it is known that weight, also a force from gravity, is identified in the equation $F_g = mg$, these two equations can be set equal to each other. The m_1 in the weight equation will cancel the m_1 value in the Universal Gravitation equation:

$$m_1 g = G\frac{m_1 m_2}{r^2}$$

$$g = G\frac{m_2}{r^2}$$

where:
m_1 = mass of an object
m_2 = mass of the planet
g = acceleration due to gravity on the planet

The equation above can be used to determine the acceleration due to gravity on or near any planet or body, given its mass and the distance from its center. The value of g in this context is often described as the **gravitational field.** Directly substituting the units into the equation above yields the unit for g as N/kg. It is already known that g also has the unit of m/s^2. Therefore, 1 N/kg = 1 m/s^2. This image below shows the gravitational field lines (force vectors) of the Earth:

The relative strength of the gravitational field can be identified in the diagram above. The closer the field lines are to each other the stronger the gravitational field. There is a stronger field (and therefore a stronger force exerted if another object were there) closer to the Earth than farther away. In comparing multiple field diagrams, the diagram with more field lines drawn has a stronger gravitational field than one with fewer field lines.

Solved Example Problems

1. The mass of Earth is 5.97×10^{24} kg, the mass of the moon is 7.35×10^{22} kg, and the mean distance of the Moon from the center of the Earth is 3.84×10^{5} km. Use this data to calculate the magnitude of the gravitational force exerted by the Earth on the Moon.

$$F_g = G\frac{m_1 m_2}{r^2}$$

$$F_g = 6.67 \times 10^{-11}\,\mathrm{Nm^2/kg^2}\left(\frac{(5.97 \times 10^{24}\,\mathrm{kg})(7.35 \times 10^{22}\,\mathrm{kg})}{(3.84 \times 10^{5}\,\mathrm{m})^2}\right)$$

$$F_g = 1.98 \times 10^{26}\,\mathrm{N}$$

2. The asteroid Vesta has a mass of 3.0×10^{20} kg and an average radius of 510 km.
   a. What is the acceleration of gravity at its surface?

   $$g = G\frac{m}{r^2}$$

   $$g = 6.67 \times 10^{-11}\,\mathrm{Nm^2/kg^2}\left(\frac{3.0 \times 10^{20}\,\mathrm{kg}}{(5.0 \times 10^{5}\,\mathrm{m})^2}\right)$$

   $$g = 8.0 \times 10^{-2}\,\frac{\mathrm{m}}{\mathrm{s^2}}$$

   b. How much would a 95 kg astronaut weigh at the surface of Vesta?

   $$F_g = mg$$

   $$F_g = (95\,\mathrm{kg})\left(8.0 \times 10^{-2}\,\frac{\mathrm{m}}{\mathrm{s^2}}\right)$$

   $$F_g = 7.6\,\mathrm{N}$$

Apparent Weight

The weight calculations above are for a stationary object on the Earth's surface. Sometimes, however, people may observe a feeling of an *apparent* weight. This often occurs while riding on amusement park rides or in an elevator. When a person is in an accelerated state in an elevator, he may feel pushed into the floor in certain movements, or lifted up in others. This is due to the fact that there is an unbalanced force acting on the person when the elevator is accelerating. A classic problem in physics, the elevator problem, demonstrates apparent weight calculations.

The Elevator Problem

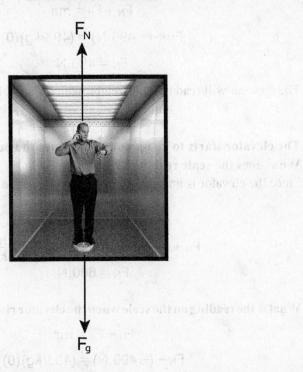

F_N

F_g

- **A person weighing 490. N stands on a scale in an elevator.**
  Before doing any calculations, it is important to determine the passenger's mass:

$$F_g = mg$$
$$-490.\,N = m\left(-9.81\frac{m}{s^2}\right)$$
$$m = 49.9\,kg$$

Since both weight (F_g) and g are directed downward they are negative.
In the diagram above, the net force is the vector sum of the two forces on him: the force of gravity (his weight), and the Normal Force (the force read from the scale). Substitute the sum for the net force in Newton's Second Law:

$$F_{net} = ma$$
$$F_N + F_g = ma$$

We can use the equation above for all parts of this problem. The value of F_N is the reading on the scale.

a. What does the scale read when the elevator is at rest?

$$F_N + F_g = ma$$

$$F_N + (-490.\,N) = (49.9\,kg)(0)$$

$$F_N = 490.\,N$$

The elevator will read normal weight, which is 490. N, since the net force on him is 0.

b. The elevator starts to go up and accelerates the person at +2.2 m/s².
What does the scale read now?
Since the elevator is now accelerating, there must be a net force on him.

$$F_N + F_g = ma$$

$$F_N + (-490.\,N) = (49.9\,kg)\left(2.2\frac{m}{s^2}\right)$$

$$F_N = 600.\,N$$

c. What is the reading on the scale when the elevator rises at a constant velocity?

$$F_N + F_g = ma$$

$$F_N + (-490.\,N) = (49.9\,kg)(0)$$

$$F_N = 490.\,N$$

The elevator will again read normal weight, which is 490. N, since the net force on him is 0.

d. The elevator slows down at –2.2 m/s² as it reaches the proper floor.
What does the scale read?

$$F_N + F_g = ma$$

$$F_N + (-490.\,N) = (49.9\,kg)\left(-2.2\frac{m}{s^2}\right)$$

$$F_N = 380.\,N$$

e. The elevator descends, accelerating at –2.7 m/s². What does the scale read?

$$F_N + F_g = ma$$

$$F_N + (-490.\,N) = (49.9\,kg)\left(-2.7\frac{m}{s^2}\right)$$

$$F_N = 355.\,N$$

f. What does the scale read with the elevator descending at a constant velocity?

$$F_N + F_g = ma$$

$$F_N + (-490.N) = (49.9\,kg)(0)$$

$$F_N = 490.N$$

g. Suppose that the cable snapped and the elevator fell freely (this is not likely to happen in real life due to redundant safety mechanisms). What would the scale read?

$$F_N + F_g = ma$$

$$F_N + (-490.N) = (49.9\,kg)\left(-9.81\,\frac{m}{s^2}\right)$$

$$F_N = 0\,N$$

Friction

Friction is a force that opposes the motion of two surfaces that are in direct contact. The frictional force is parallel to the surfaces in contact and opposite in direction to the motion. There are two main types of friction: kinetic (moving) friction, and static (stationary) friction. The static frictional force on an object is larger than the kinetic frictional force. This results in it being more difficult to start an object moving from rest as compared to keeping an object moving. There are multiple types of kinetic friction. These include sliding, fluid (friction from liquids or gases), and rolling (from wheels or bearings). Listed in order from smallest amount of frictional force to largest: fluid, rolling, sliding, and static friction.

The force of friction depends on two factors: the Normal Force (F_N) acting on the object and the **coefficient of friction** (μ). The Normal Force is the force of support from a surface such as a table or floor. Note that the Normal Force is only equal to the weight of the object if it is on a level horizontal surface. If the object is on an incline, the Normal Force is not equal to the weight, as illustrated on Page 47. The coefficient of friction is a quantity that is a property of the types of surfaces that are

Approximate Coefficients of Friction		
	Kinetic	Static
Rubber on concrete (dry)	0.68	0.90
Rubber on concrete (wet)	0.58	
Rubber on asphalt (dry)	0.67	0.85
Rubber on asphalt (wet)	0.53	
Rubber on ice	0.15	
Waxed ski on snow	0.05	0.14
Wood on wood	0.30	0.42
Steel on steel	0.57	0.74
Copper on steel	0.36	0.53
Teflon on Teflon	0.04	

sliding over each other. It does not have a unit. Some selected coefficients of friction are listed above in the chart from the Physics Reference Tables.

It is important to note that the coefficient of friction is not a force; it is a ratio of the Force of Friction (F_f) and the Normal Force (F_N). Notice that the units of the forces cancel and therefore μ has no units. It can be expressed in equation form as follows:

$$\mu = \frac{F_f}{F_N}$$

When solved for the Force of Friction:

$$F_f = \mu F_N$$

Upon inspection of the above equation, you will notice that there are only two factors that affect the amount of frictional force on an object: the Normal Force and the coefficient of friction. There are factors that *do not affect the force of friction* that are often mistaken as factors that do, such as how fast an object is moving and its contact surface area.

Solved Practice Problems

1. A smooth wooden block is placed on a smooth wooden tabletop. You find that you must exert a force of 14.0 N to keep the 40.0 N block moving at a constant velocity.
   a. What is the coefficient of sliding friction for the block and the table?

$$\mu = \frac{F_f}{F_N}$$
$$\mu = \frac{14.0\,\text{N}}{40.0\,\text{N}}$$
$$\mu = 0.35$$

   b. If a 20.0 N brick is placed on the block, what force will be required to keep the block and brick moving at a constant velocity?

$$F_f = \mu F_N$$
$$F_f = (0.35)(60.0\,\text{N})$$
$$F_f = 21\text{N}$$

2. A man is trying to push a 200.0 kg chest of drawers at rest on the floor. The coefficient of static friction between the floor and the chest of drawers is 0.45. How much horizontal force must he exert on the chest of drawers to move it?

$$F_f = \mu F_N$$
$$F_f = (0.45)(200.0\text{kg})\left(9.81\,\frac{\text{m}}{\text{s}^2}\right)$$
$$F_f = 880\text{N}$$

Ramps and Inclined Planes

All of the concepts addressed in this unit can be applied to an object on an inclined plane as well as a horizontal surface. An inclined plane is any surface that is at an angle, θ, with respect to the horizontal. An example of an object on an inclined plane is a car parked on a hilly street that makes a 30° incline with the horizontal. A free-body diagram of this car will look like the diagram below:

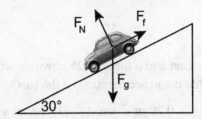

In the diagram, F_g is the weight of the car. This can be found by using the equation $F_g = mg$. Note that the Normal Force (F_N) and the frictional force (F_f) are perpendicular to each other, but not to the weight. In order to properly solve this situation, the *components of the weight vector* must be found. These components are perpendicular to the incline ($F_{g\perp}$) and parallel to the incline ($F_{g\parallel}$). They will look like this:

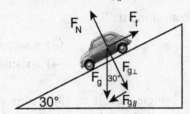

In the drawing above, the angle between the weight vector (F_g) and the perpendicular component of the weight ($F_{g\perp}$) is the same as the angle of inclination of the ramp. Using right triangle trigonometry, the equations for finding $F_{g\perp}$ and $F_{g\parallel}$ are:

$$F_{g\parallel} = F_g \sin\theta$$
$$F_{g\perp} = F_g \cos\theta$$

If the object is at rest, the acceleration is zero, and it is in equilibrium. Therefore, the force of friction is equal in magnitude to $F_{g\parallel}$. If the object is accelerating, the net force is the vector sum of these two forces. The normal force will always be equal to $F_{g\perp}$ for an object on an incline.

Unit 3
Practice Questions

1. A Force of 25 newtons east and a force of 25 newtons west act concurrently on a 5.0-kilogram cart. What is the acceleration of the cart?

   (1) 1.0 m/s² west   (2) 0.20 m/s² east   (3) 5.0 m/s² east   (4) 0 m/s²

2. What is the acceleration due to gravity at a location where a 15.0-kilogram mass weighs 45.0 newtons?

   (1) 675 m/s²   (2) 9.81 m/s²   (3) 3.00 m/s²   (4) 0.333 m/s²

3. Which object has the greatest inertia?

   (1) a falling leaf   (3) a seated high school student
   (2) a softball in flight   (4) a rising helium-filled toy balloon

4. A 1200-kilogram space vehicle travels at 4.8 meters per second along the level surface of Mars. If the magnitude of the gravitational field strength on the surface of Mars is 3.7 newtons per kilogram, the magnitude of the normal force acting on the vehicle is

   (1) 320 N   (2) 930 N   (3) 4400 N   (4) 5800 N

5. An 80-kilogram skier slides on waxed skis along a horizontal surface of snow at constant velocity while pushing with his poles. What is the horizontal component of the force pushing him forward?

   (1) 0.05 N   (2) 0.4 N   (3) 40 N   (4) 4 N

6. As a meteor moves from a distance of 16 Earth radii to a distance of 2 Earth radii from the center of Earth, the magnitude of the gravitational force between the meteor and Earth becomes

   (1) 1/8 as great   (3) 64 times as great
   (2) 8 times as great   (4) 4 times as great

7. A block weighing 10.0 newtons is on a ramp inclined at 30.0° to the horizontal. A 3.0-newton force of friction, F_f, acts on the block as it is pulled up the ramp at constant velocity with force F, which is parallel to the ramp, as shown in the diagram below.

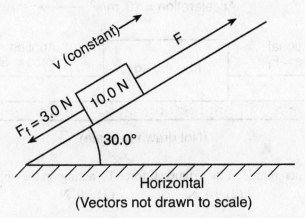

Horizontal
(Vectors not drawn to scale)

What is the magnitude of force F?

(1) 7.0 N   (2) 8.0 N   (3) 10. N   (4) 13 N

8. Earth's mass is approximately 81 times the mass of the Moon. If Earth exerts a gravitational force of magnitude F on the Moon, the magnitude of the gravitational force of the Moon on Earth is

(1) F   (2) $\dfrac{F}{81}$ (3) 9F   (4) 81F

9. The diagram shows two bowling balls, A and B, each having a mass of 7.00 kilograms, placed 2.00 meters apart.

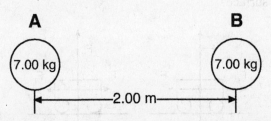

What is the magnitude of the gravitational force exerted by ball A on ball B?

(1) 8.17×10^{-9} N   (3) 8.17×10^{-10} N

(2) 1.63×10^{-9} N   (4) 1.17×10^{-10} N

10. The diagram below shows a 4.0-kilogram object accelerating at 10. meters per second² on a rough horizontal surface.

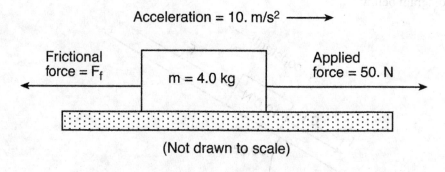

Acceleration = 10. m/s² ⟶

Frictional force = F_f

m = 4.0 kg

Applied force = 50. N

(Not drawn to scale)

What is the magnitude of the frictional force F_f acting on the object?
(1) 5.0 N (2) 10. N (3) 20. N (4) 40. N

11. Which diagram best represents the gravitational field lines surrounding Earth?

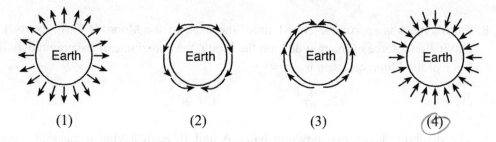

Earth   Earth   Earth   Earth

(1)   (2)   (3)   (4)

12. Which vector diagram best represents a cart slowing down as it travels to the right on a horizontal surface?

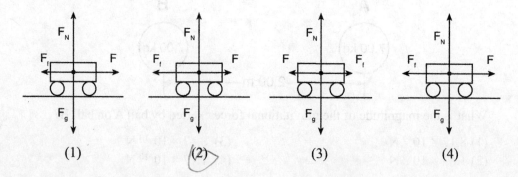

(1)   (2)   (3)   (4)

13. The diagram below shows a 5.00-kilogram block at rest on a horizontal, frictionless table.

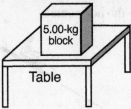

Which diagram best represents the force exerted on the block by the table?

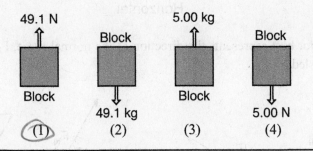

14. The diagram below represents a block at rest on an incline.

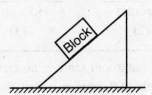

Which diagram best represents the forces acting on the block? (F_f = frictional force, F_N = normal force, and F_w = weight.)

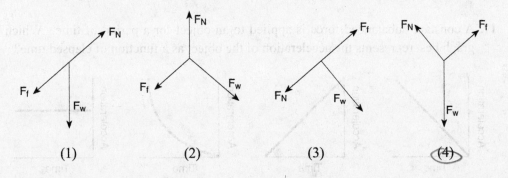

15. The diagram below shows a sled and rider sliding down a snow-covered hill that makes an angle of 30.° with the horizontal.

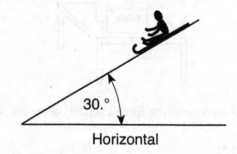

Which vector best represents the direction of the normal force, F$_N$, exerted by the hill on the sled?

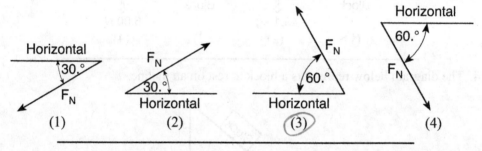

16. A person is standing on a bathroom scale in an elevator car. If the scale reads a value greater than the weight of the person at rest, the elevator car could be moving

(1) downward at constant speed   (3) downward at increasing speed
(2) upward at constant speed   (4) upward at increasing speed

17. A constant unbalanced force is applied to an object for a period of time. Which graph best represents the acceleration of the object as a function of elapsed time?

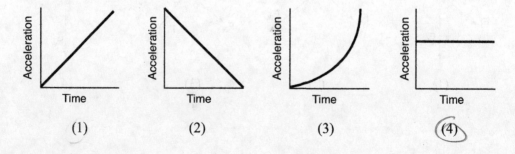

18. A box is pushed toward the right across a classroom floor. The force of friction on the box is directed toward the

(1) left   (2) right   (3) ceiling   (4) floor

19. A 60-kilogram skydiver is falling at a constant speed near the surface of Earth. The magnitude of the force of air friction acting on the skydiver is approximately

(1) 0 N   (2) 6 N   (3) 60 N   (4) 600 N

20. When a 12-newton horizontal force is applied to a box on a horizontal tabletop, the box remains at rest. The force of static friction acting on the box is

(1) 0 N   (3) 12 N
(2) between 0 N and 12 N   (4) greater than 12 N

21. A 1.5-kilogram lab cart is accelerated uniformly from rest to a speed of 2.0 meters per second in 0.50 second. What is the magnitude of the force producing this acceleration?

(1) 0.70 N   (2) 1.5 N   (3) 3.0 N   (4) 6.0 N

22. Which graph best represents the motion of an object that is *not* in equilibrium as it travels along a straight line?

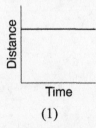 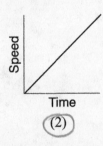 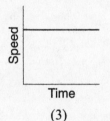 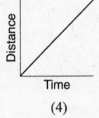

(1)   (2)   (3)   (4)

23. Three forces act on a box on an inclined plane as shown in the diagram below. [Vectors are not drawn to scale.]

If the box is at rest, the net force acting on it is equal to

(1) the weight
(2) the normal force
(3) friction
(4) zero

24. A net force of 10. newtons accelerates an object at 5.0 meters per second^2. What net force would be required to accelerate the same object at 1.0 meter per second^2?

(1) 1.0 N   (2) 2.0 N   (3) 5.0 N   (4) 50. N

25. The diagram below shows a force of magnitude F applied to a mass at angle θ relative to a horizontal frictionless surface.

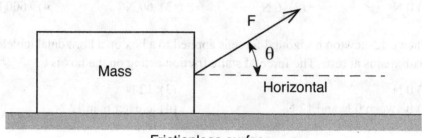

As angle θ is increased, the horizontal acceleration of the mass

(1) decreases   (2) increases   (3) remains the same

UNIT 4
Motion in Two Dimensions - Circles and Projectiles

Up to this point, all analysis of the motion of objects was restricted to motion in one dimension. That means the objects are only traveling in a straight line. In this unit, we will analyze the motion of objects that are traveling in two dimensions. First will be an analysis of motion in a circle, followed by an analysis of projectile motion.

Uniform Circular Motion

Newton's First Law of Motion, the Law of Inertia, states that an object will continue in a straight line at constant speed unless there is an unbalanced force acting on it. With that understanding, in order for an object to travel in a circle, there must then be an unbalanced force acting on the object to change its direction.

We will consider objects traveling in a circular path at constant speed only. This type of motion is called **Uniform Circular Motion.**

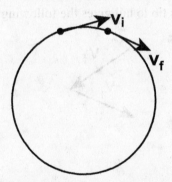

In the diagram above, the velocity vector for an object traveling in a clockwise circle is shown for two different positions. Notice that the length of the vector is the same for both positions, but it is pointing in a different direction at each position. The magnitude of the velocity has not changed, but the direction has. So, is this object accelerating? Yes, since the definition of acceleration is the rate of change of *velocity*, it must be accelerating. However, the equation previously used for acceleration will not apply here, as the magnitude of the velocity is not changing. We must take a different approach to determine this acceleration.

A change in velocity is typically expressed as Δv, since the Greek letter *delta* refers to a change. It can also be expressed as v_f-v_i. Since the magnitudes of the velocities are not changing, we must look at the change in the direction of the velocity vectors. This requires **vector subtraction.** Technically speaking, two vectors can't be subtracted.

In order to perform this operation, you can add the negative of the second vector to the first. A negative vector is the same as the original vector, except it points in exactly the opposite direction. So, in terms of vectors, we have:

$$v_f - v_i = v_f + (-v_i)$$

The velocity vectors from the diagram above will then appear as follows:

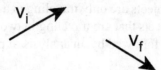

In order to subtract these vectors, we must determine $-v_i$. This vector will be exactly the same as v_i in magnitude, but will point in the opposite direction.

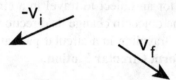

Rearranging these vectors tip to tail gives the following:

The resultant will then be:

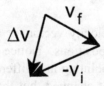

As can be seen in the diagram above, the direction of the Δv vector is toward the center of the circle. Therefore, the centripetal acceleration is directed toward the center of the circular path. To determine the magnitude of the centripetal acceleration, the following equation is used:

$$a_c = \frac{v^2}{r}$$

Centripetal Force

From Newton's Laws of Motion, we know that in order for an object to accelerate, there must be a net force acting upon the object. In the case of uniform circular motion, this force is changing the direction of the object. According to Newton's Second Law:

$$F_{net} = ma$$

The acceleration in the case of circular motion has a special name: **centripetal acceleration** (NOT centrifugal or centrifical!!!). The word centripetal means center-seeking, or directed toward the center. As a result, Newton's Second Law can be redefined for circular motion as:

$$F_c = ma_c$$

where:
F_c = Centripetal Force
m = mass
a_c = centripetal acceleration

Knowing the equation for centripetal acceleration from the previous section, the formula for centripetal force can also be expressed as:

$$F_c = m\frac{v^2}{r}$$

What is the Centripetal Force? What is providing it? Imagine an experiment where you whirl an object tied to a string in a circle over your head, as in the diagram below.

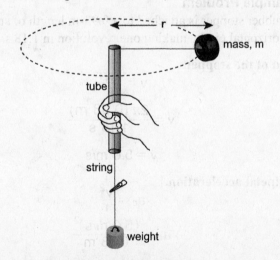

Centripetal Force Apparatus

In this experiment, the force to keep the mass moving in a circle is provided by the string. Specifically, the string along the radius, r, of the circular path is providing the force. The direction of this force is toward the center of the circular path. If this force were to suddenly disappear, what would the path of the object be? Without this force, there would no longer be an unbalanced force on the object so it would then travel in a straight line at constant speed. Relative to the circular path, this would be *tangent to the circle.* To review, the direction of the centripetal force vector is toward the center of the circle, and the direction of the velocity vector is tangent to the circle, as in the diagram below.

Circular Speed

The speed of any object is found by dividing the distance traveled by the time it takes to travel that distance. In kinematics, the equation was:

$$v = \frac{d}{t}$$

In the case of an object traveling in a circle, the distance traveled is the *circumference of the circle.* From geometry, the circumference of a circle is found by πd (pi times the diameter), or 2πr (2 times pi times the radius). As a result, the equation for calculating the speed of an object traveling in a circle can be rewritten as:

$$v = \frac{2\pi r}{t} \qquad \text{or} \qquad v = \frac{\pi d}{t}$$

Solved Example Problem

A 0.013-kg rubber stopper is attached to a 0.93-m length of string. The stopper is swung in a horizontal circle, making one revolution in 1.18 s. Determine:

a. the speed of the stopper.

$$v = \frac{2\pi r}{t}$$

$$v = \frac{2\pi (0.93 \text{ m})}{1.18 \text{ s}}$$

$$v = 5.0 \text{ m/s}$$

b. its centripetal acceleration.

$$a_c = \frac{v^2}{r}$$

$$a_c = \frac{(5.0 \text{ m/s})^2}{0.93 \text{ m}}$$

$$a_c = 27 \text{ m/s}^2$$

c. the force the string exerts on the rubber stopper.

$$F_c = ma_c$$

$$F_c = (0.013 \text{ kg})(27 \text{ m/s}^2)$$

$$F_c = 0.35 \text{ N}$$

CIRCLES AND PROJECTILES

Projectile Motion

A projectile is any object that is thrown, fired, kicked, batted, etc. through the air so that it travels both in the horizontal (x) and vertical (y) directions simultaneously. The general shape of a projectile's path is that of a parabola. To simplify the analysis of projectile motion, the effects of air resistance will be ignored. In addition, we will consider only horizontally launched projectiles and those launched at an angle that land at the same elevation as the launch.

The equations of kinematics will apply to projectile motion. There is one very important factor, however. In order to analyze projectile motion, *the horizontal, or x, motion and the vertical, or y, motion must be separated when substituting into equations.* That means that any numbers, or data, pertaining to the horizontal motion will never appear in an equation that contains data pertaining to the vertical direction. We must take the existing kinematics expressions and customize them for the x and y directions.

The equations to be customized include the following:

$$d = v_i t + \tfrac{1}{2} a t^2$$
$$v_f^2 = v_i^2 + 2ad$$
$$v_f = v_i + at$$

When customizing these equations for the x and y directions, imagine the actual movement of a projectile in motion. Imagine a soccer ball that is kicked so that it leaves the ground as it travels. That soccer ball is moving forward as it also moves in the vertical direction, first up to its highest point, and then downward as it returns to the ground. The entire time, the soccer ball is moving forward until it hits the ground and eventually stops. Considering *just the forward direction*, there is nothing (unless air resistance is to be considered) that would cause it to speed up or slow down horizontally. Therefore, <u>the horizontal acceleration is zero</u> and the horizontal velocity is constant. Now, considering *just the vertical direction of travel*, gravity will cause the object to have an acceleration. Since gravitational acceleration is directed toward the center of the earth, <u>gravity only acts in the vertical direction and has no effect on the horizontal motion</u>.

When customizing the kinematics equations for the horizontal and vertical components of the projectile's motion, it is helpful to separate these equations and associated data into a table. The resulting equations appear below:

Original Equation	Horizontal (x) Direction	Vertical (y) Direction
$d = v_i t + \tfrac{1}{2} a t^2$	$d_x = v_{ix} t$	$d_y = v_{iy} t + \tfrac{1}{2} g t^2$
$v_f^2 = v_i^2 + 2ad$	N/A	$v_{fy}^2 = v_{iy}^2 + 2g d_y$
$v_f = v_i + at$	N/A	$v_{fy} = v_{iy} + gt$

where:

d_x = horizontal distance (range)

v_{ix} = initial horizontal velocity

t = time

d_y = vertical distance (height)

v_{iy} = initial vertical velocity

v_{fy} = final vertical velocity

g = acceleration due to gravity

(9.81 m/s^2, directed down)

The sign of the velocity is important in indicating the direction of travel in two-dimensional kinematics problems, as it is in one-dimensional kinematics problems. Typically, anything to the right is positive for the x direction, and to the left is negative. In the vertical direction, *up* is positive and *down* is negative. As a result, the acceleration due to gravity, g, will be -9.81 m/s² (regardless of which way the object is moving).

If the projectile's vertical component of the velocity is up, it has a positive v_y; if the vertical component of the velocity is down, then it has a negative v_y. For a projectile launched at an angle, during the first half of flight there will be a positive vertical velocity and a negative vertical acceleration. Gravity causes the projectile to slow down while it is moving upward in the vertical direction. For the second half of that projectile's flight, both velocity and acceleration in the vertical direction are down (negative) as it falls. Gravity causes it to speed up in the vertical direction during this time.

Horizontally Launched Projectiles

A horizontally launched projectile is any object that is projected straight out from the edge of a cliff, table, or other surface and then travels forward and downward toward the landing point. An example of this would be a marble rolled off the edge of a table and allowed to fall to the floor.

In this example, the initial vertical velocity (v_{iy}) is zero. The initial horizontal velocity is the speed at which it leaves the edge of the table. Once it leaves the edge of the table, the vertical velocity increases (from zero) and the horizontal velocity remains constant. The individual velocity vectors for the horizontal and vertical directions at various positions are shown below. Notice that the v_y is steadily increasing and the v_x vector is remaining constant.

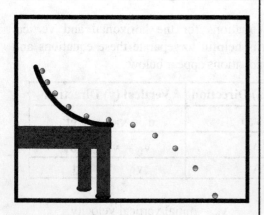

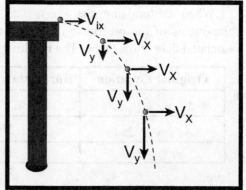

CIRCLES AND PROJECTILES

One common laboratory experiment dealing with horizontal projectiles is often nicknamed "The Projectile Contest." In this experiment, one must roll a marble down a ramp at an unknown height and predict where it will land by placing a cup at that location.

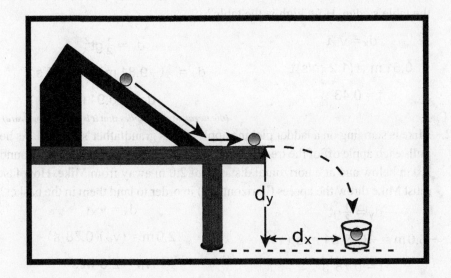

By performing experiments at a known height, the range (d_x), and height (d_y) can be directly measured. From these measurements, the time in the air can be calculated by using the following equation:

$$d_y = v_{iy}t + \tfrac{1}{2}gt^2$$

Since this is a horizontally launched projectile, v_{iy} is zero, so the expression becomes:

$$d_y = \tfrac{1}{2}gt^2$$

In this equation, since the marble is falling downward, both d_y and g are negative. The time can be calculated, and since time is a scalar quantity, it can be used in both equations for x and y motion. Once the time is calculated, it can be used to find the horizontal velocity (v_x) using:

$$d_x = v_{ix}t$$

When the ramp is at a new, unknown height, what changes and what remains the same? At this new height the value of d_y is obviously different, the time in the air and range will be different, but, the horizontal velocity will be the same. Knowing this, the new height can be measured, the time and range can be calculated, and the cup can be placed at this predicted location.

Solved Example Problems

1. In Physics Lab, John rolls a 10 g marble down a ramp and off of a table with a horizontal velocity of 1.2 m/s. The marble falls in a cup placed 0.51 m from the table's edge. How high is the table?

$$d_x = v_{ix}t$$

$$0.51 \text{ m} = (1.2 \text{ m/s})t$$

$$t = 0.43 \text{ s}$$

$$d_y = \tfrac{1}{2}gt^2$$

$$d_y = \tfrac{1}{2}(-9.81 \text{ m/s}^2)(0.43 \text{ s})^2$$

$$d_y = -0.91 \text{ m}$$

(the negative sign implies that it is falling downward)

2. Mike is standing on a ladder picking apples in his grandfather's orchard. As he pulls each apple off of the tree, he tosses it into a basket that sits on the ground 3.0 m below and at a horizontal distance of 2.0 m away from Mike. How fast must Mike throw the apples (horizontally) in order to land them in the basket?

$$d_y = \tfrac{1}{2}gt^2$$

$$-3.0 \text{m} = \tfrac{1}{2}(-9.81 \text{ m/s})(t)^2$$

$$t = 0.78 \text{ s}$$

$$d_x = v_{ix}t$$

$$2.0 \text{m} = (v_{ix})(0.78 \text{ s})$$

$$v_{ix} = 2.6 \text{ m/s}$$

Projectiles Launched at an Angle

Projectiles that are launched at an angle will be considered only when they take off from and land at the same elevation. An example is in the diagram below:

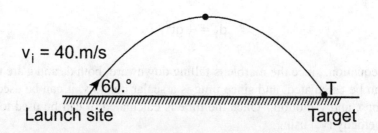

$v_i = 40.\text{m/s}$ 60.°   Launch site   T   Target

In this example, the angle of launch (60.°) and the initial launch velocity (40. m/s) are provided. Note that this initial velocity is in neither the x direction nor the y direction. This is the initial velocity at an angle. Up to this point, all equations separated the x and y motions and did not utilize any motion at an angle. In order to solve a projectile problem like this, the first thing to do is to determine the x and y components of this initial velocity. This is best done using some right-triangle trigonometry.

CIRCLES AND PROJECTILES

The initial velocity of a projectile v_i can be resolved into components in the x (v_{ix}) and y (v_{iy}) direction as shown:

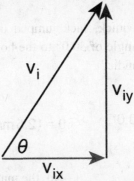

Using the properties of a right triangle, v_i is the hypotenuse, v_{ix} is the leg adjacent to the angle θ, and v_{iy} is the leg opposite the angle θ. With the mnenonic SOHCAHTOA, expressions can be written based on these parts of the triangle:

$$\sin\theta = \frac{v_{iy}}{v_i} \qquad \text{or} \qquad v_{iy} = v_i \sin\theta$$

$$\cos\theta = \frac{v_{ix}}{v_i} \qquad \text{or} \qquad v_{ix} = v_i \cos\theta$$

The previous equations are similar to a pair of equations from the Reference Tables, as listed below:

$$A_y = A \sin\theta$$
$$A_x = A \cos\theta$$

The equations from the Reference Tables refer to A as any vector quantity. For the purposes of projectiles, A can refer to the initial velocity, v_i, and then the equations are exactly the same.

Some things to consider with projectiles:

- The time it takes to reach the highest point in its trajectory (path) is exactly half of the total time of flight.

- The vertical velocity (v_y) at the highest point of its trajectory is zero.

- The vertical velocity at launch is equal in magnitude but opposite in direction to the vertical velocity at landing.

- The horizontal velocity is constant.

- The (only) acceleration of the projectile is that caused by gravity ($g = 9.81$ m/s^2 straight down).

Solved Example Problems

1. Jack be nimble, Jack be quick, Jack jumped over the candlestick with a velocity of 5.0 m/s at an angle of 30.0° to the horizontal. Did Jack burn his feet on the 0.25-m-high candle?

$$v_{iy} = v_i \sin \theta$$

$$v_{iy} = (5.0 \text{ m/s}) \sin 30.0°$$

$$v_{iy} = 2.5 \text{ m/s}$$

$$v_{fy}^2 = v_{iy}^2 + 2gd_y$$

$$0 = (2.5 \text{ m/s})^2 + 2(-9.81 \text{ m/s}^2 d_y)$$

$$d_y = 0.32 \text{ m}$$

No, the maximum height of the jump is higher than the candle.

2. Courtney kicks a soccer ball initially at rest on level ground giving it an initial velocity of 7.8 m/s at an angle of 32°.

   a. How long will the ball be in the air?

$$v_{iy} = v_i \sin \theta$$

$$v_{iy} = (7.8 \text{ m/s}) \sin 32°$$

$$v_{iy} = 4.1 \text{ m/s}$$

$$d_y = v_{iy}t + \tfrac{1}{2}gt^2$$

$$0 = (4.1 \text{ m/s})t + \tfrac{1}{2}(-9.81 \text{ m/s}^2)t^2$$

$$t = 0.84 \text{ s}$$

   b. What is the maximum height of the ball?

$$v_{fy}^2 = v_{iy}^2 + 2gd_y$$

$$0 = (4.1 \text{ m/s})^2 + 2(-9.81 \text{ m/s}^2)d_y$$

$$d_y = 0.86 \text{ m}$$

   c. What will be its range?

$$v_{ix} = v_i \cos \theta$$

$$v_{ix} = (7.8 \text{ m/s}) \cos 32°$$

$$v_{ix} = 6.6 \text{ m/s}$$

$$d_x = v_{ix}t$$

$$d_x = (6.6 \text{ m/s})(0.84 \text{ s})$$

$$d_x = 5.5 \text{ m}$$

Unit 4
Practice Questions

1. The diagram below represents the path of a stunt car that is driven off a cliff, neglecting friction.

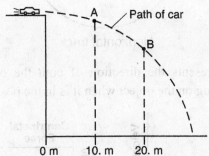

Compared to the horizontal component of the car's velocity at point A, the horizontal component of the car's velocity at point B is

(1) smaller   (2) greater   (3) the same

2. Two stones, A and B, are thrown horizontally from the top of a cliff. Stone A has an initial speed of 15 meters per second and stone B has an initial speed of 30. meters per second. Compared to the time it takes stone A to reach the ground, the time it takes stone B to reach the ground is

(1) the same   (3) half as great
(2) twice as great   (4) four times as great

3. A 1750-kilogram car travels at a constant speed of 15.0 meters per second around a horizontal, circular track with a radius of 45.0 meters. The magnitude of the centripetal force acting on the car is

(1) 5.00 N   (2) 583 N   (3) 8750 N   (4) 3.94×10^5 N

4. A 0.50-kilogram object moves in a horizontal circular path with a radius of 0.25 meter at a constant speed of 4.0 meters per second. What is the magnitude of the object's acceleration?

(1) 8.0 m/s^2 (2) 16 m/s^2 (3) 32 m/s^2 (4) 64 m/s^2

5. The diagram below shows an object moving counterclockwise around a horizontal, circular track.

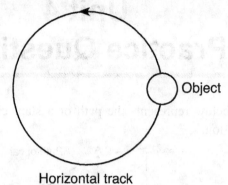

Horizontal track

Which diagram represents the direction of both the object's velocity and the centripetal force acting on the object when it is in the position shown?

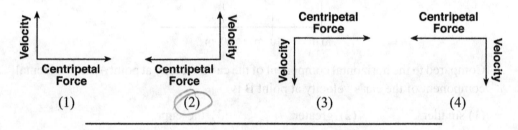

6. Which graph best represents the relationship between the magnitude of the centripetal acceleration and the speed of an object moving in a circle of constant radius?

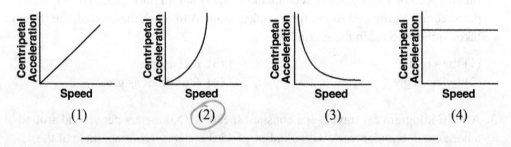

7. A machine launches a tennis ball at an angle of 25° above the horizontal at a speed of 14 meters per second. The ball returns to level ground. Which combination of changes *must* produce an increase in time of flight of a second launch?

(1) decrease the launch angle and decrease the ball's initial speed
(2) decrease the launch angle and increase the ball's initial speed
(3) increase the launch angle and decrease the ball's initial speed
(4) increase the launch angle and increase the ball's initial speed

CIRCLES AND PROJECTILES

8. A golf ball is propelled with an initial velocity of 60. meters per second at 37° above the horizontal. The horizontal component of the golf ball's initial velocity is

(1) 30. m/s   (2) 36 m/s   (3) 40. m/s   (4) 48 m/s

9. A volleyball hit into the air has an initial speed of 10. meters per second. Which vector best represents the angle above the horizontal that the ball should be hit to remain in the air for the greatest amount of time?

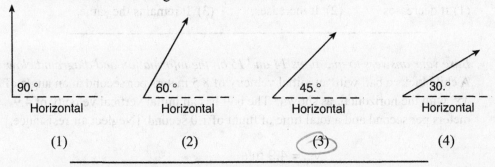

90.°   60.°   45.°   30.°

Horizontal   Horizontal   Horizontal   Horizontal

(1)   (2)   (3)   (4)

---

10. A golf ball is hit at an angle of 45° above the horizontal. What is the acceleration of the golf ball at the highest point in its trajectory? [Neglect friction.]

(1) 9.8 m/s² upward   (3) 6.9 m/s² horizontal
(2) 9.8 m/s² downward   (4) 0.0 m/s²

11. A ball is thrown horizontally at a speed of 24 meters per second from the top of a cliff. If the ball hits the ground 4.0 seconds later, approximately how high is the cliff?

(1) 6.0 m   (2) 39 m   (3) 78 m   (4) 96 m

12. In the diagram below, S is a point on a car tire rotating at a constant rate.

Which graph best represents the magnitude of the centripetal acceleration of point S as a function of time?

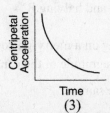

Centripetal Acceleration   Centripetal Acceleration   Centripetal Acceleration   Centripetal Acceleration

Time   Time   Time   Time
(1)   (2)   (3)   (4)

13. The diagram below represents the path of an object after it was thrown.

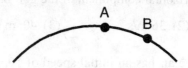

What happens to the object's acceleration as it travels from A to B? [Neglect friction]

(1) It decreases.   (2) It increases.   (3) It remains the same.

Base your answers to questions 14 and 15 on the information and diagram below.
A child kicks a ball with an initial velocity of 8.5 meters per second at an angle of
35° with the horizontal, as shown. The ball has an initial vertical velocity of 4.9
meters per second and a total time of flight of 1.0 second. [Neglect air resistance.]

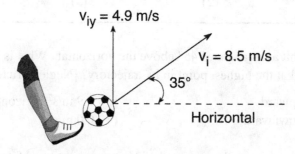

14. The horizontal component of the ball's initial velocity is approximately

(1) 3.6 m/s   (2) 4.9 m/s   (3) 7.0 m/s   (4) 13 m/s

15. The maximum height reached by the ball is approximately
(1) 1.2 m   (2) 2.5 m   (3) 4.9 m   (4) 8.5 m

---

16. A ball of mass M at the end of a string is swung in a horizontal circular path of
radius R at constant speed V. Which combination of changes would require the
greatest increase in the centripetal force acting on the ball?

(1) doubling V and doubling R   (3) halving V and doubling R
(2) doubling V and halving R   (4) halving V and halving R

17. A child is riding on a merry-go-round. As the speed of the merry-go-round is
doubled, the magnitude of the centripetal force acting on the child

(1) remains the same   (3) is halved
(2) is doubled   (4) is quadrupled

68

18. An archer uses a bow to fire two similar arrows with the same string force. One arrow is fired at an angle of 60.° with the horizontal, and the other is fired at an angle of 45° with the horizontal. Compared to the arrow fired at 60.°, the arrow fired at 45° has a

(1) longer flight time and longer horizontal range
(2) longer flight time and shorter horizontal range
(3) shorter flight time and longer horizontal range
(4) shorter flight time and shorter horizontal range

19. The diagram below shows a student throwing a baseball horizontally at 25 meters per second from a cliff 45 meters above the level ground.

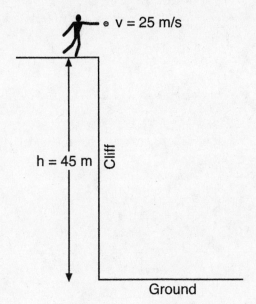

Approximately how far from the base of the cliff does the ball hit the ground? [Neglect air resistance]

(1) 45 m   (2) 75 m   (3) 140 m   (4) 230 m

---

20. A projectile is fired from a gun near the surface of Earth. The initial velocity of the projectile has a vertical component of 98 meters per second and a horizontal component of 49 meters per second. How long will it take the projectile to reach the highest point in its path?

(1) 5.0 s   (2) 10. s   (3) 20. s   (4) 100. s

UNIT 5
Momentum

Any object in motion has **momentum**. Momentum is a vector quantity that is represented by the product of an object's mass and velocity. In equation form:

$$p = mv$$

where:

p = momentum

m = mass

v = velocity

Looking directly at the units of the quantities above, mass is typically measured in kilograms (kg) and velocity is typically measured in meters per second (m/s). Therefore, the unit of momentum is kg • m/s. Considering that the Newton is equivalent to 1 kg • m/s^2, momentum can also be expressed as a N·s. The importance of this representation of the unit of momentum will be explained in the next section. The direction of an object's momentum is the same as the object's velocity.

It is important not to confuse momentum with inertia. Inertia is a property of all objects with mass, regardless of whether they are moving. Momentum is a property of objects with mass only if they are in motion. Often times there are examination questions asking which object has the greatest inertia (the one with the greatest mass), or the greatest momentum (the one with the greatest *product* of mass *and* velocity).

Changing Momentum

In order to change the momentum of an object, the mass, the velocity, or both quantities must change. In rare cases the mass of an object will change. The most common situations with a change in mass might be a rocket blasting off and burning large quantities of fuel and/or ejecting a booster engine, or a snowball rolling down a hill and getting larger as it rolls. More often than changing mass, the object will experience a change in velocity. Using this most common situation, an expression for change in momentum is:

$$\Delta p = m \Delta v$$

From the kinematics unit, it is known that acceleration is a change in velocity per unit time ($a = \dfrac{\Delta v}{t}$). Solving this expression for Δv: $\Delta v = at$. This can be substituted into the equation for change in momentum from above.

$$\Delta p = mat$$

From Newton's Second Law, $F_{net} = ma$. This can be substituted into the last expression as:

$$\Delta p = F_{net} t$$

In order to change the momentum of an object, there must be a net force exerted on that object, and that force must be exerted for a certain amount of time. This is called an **Impulse** (J) on an object. The result is the **impulse-momentum theorem**, which states that the impulse imparted on an object, which is equal to the product of the net force and the time, is equal to the change in momentum of that object. In equation form:

$$J = F_{net}t = \Delta p$$

There are many situations in real life where the impulse-momentum theorem is helpful. For many years the dashboards and steering wheels of cars have been padded a bit. If your head were to make contact with the dashboard during a collision, the impact would be slightly cushioned. But what does this mean in terms of impulse and momentum? During the impact, the speed of your head is reduced from whatever speed it was moving to zero. By padding the dashboard, it increases the time that it would take to stop your head and, therefore, the net force exerted on your head would be less. Air bags in cars further increase the time to stop your head. This decreases the net force exerted on you and, as a result, injuries and deaths have been significantly reduced since the inclusion of air bags in cars.

Imagine that you are playing catch with a water balloon. You do not want the balloon to break when you catch it. In which direction do you move your hands to safely catch the balloon: toward the balloon or with the balloon's travel direction? Do you hold your hands still? In order to reduce the force the most, you should move your hands with the balloon to increase the contact time to stop the balloon. Then, as with a car's air bag, by increasing the time of contact, there is less net force exerted. Other examples of this in action include bending your knees while landing a jump and bumpers on cars absorbing the impact during collisions.

MOMENTUM

Momentum Conservation

In any *closed, or isolated system*, the total momentum is *conserved*. An isolated system is any collection of objects in which there are no external forces present. The total momentum of all objects in this system remains constant, but the momentum may shift from one object to another. This typically can happen during a *collision* among objects. Since momentum is a vector quantity, collisions, explosions, or other events may also occur in two (or even three) dimensions. One-dimensional collisions are the most common type analyzed for momentum conservation. Expressed mathematically:

$$P_{before} = P_{after}$$

The expression above refers to the fact that the total momentum of a system of objects *before* a collision, explosion, or other event occurs is equal to the total momentum of that same system of objects *after* that event has occurred. Considering that there are often multiple objects in the system, the expression above is not really an equation, but more of a guideline for the creation of a custom equation for the specific problem being solved. Following the method below, the expression often expands to include the objects in the system and may look like this:

$$m_1v_1 + m_2v_2 = m_1v_1' + m_2v_2'$$

The equation above represents a two-object system that undergoes a collision. A type of collision after which two objects separate from each other is referred to as an **elastic collision**. Examples of this type of collision include billiard balls colliding and a Newton's Cradle device. Numerical subscripts are used to denote the separate objects involved in a collision. The superscript notation of prime (′) represents a value that exists *after* the collision or event.

If two objects stick together upon collision, the equation would look like this:

$$m_1v_1 + m_2v_2 = (m_1 + m_2)v'$$

Note that there is only one momentum after the collision, and the mass is the combined mass of both objects. This often occurs when two carts, or rail cars, collide and lock together. A type of collision after which two objects remain joined is referred to as an **inelastic collision**.

If one object explodes or splits into two individual objects, the equation would be:

$$(m_1 + m_2)v = m_1 v'_1 + m_2 v'_2$$

Note in this equation that there is one object before the event and two after. Examples of this include firing a gun or a cannon, and the splitting apart of two lab carts held together with a spring. Before the event there is one object (e.g. the gun and bullet combined), and after the event there are two separate objects. In the gun example, the bullet moves forward and the gun moves backward (recoils). The signs of the velocities would be opposite to denote the movement in opposite directions.

before

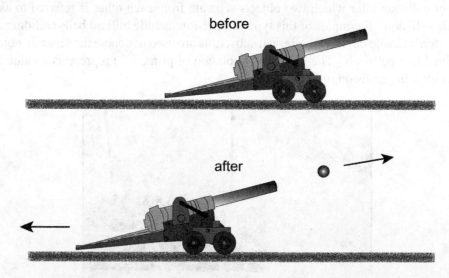

after

The specific equation used for conservation of momentum problems depends on the details of the problem being solved. Following the steps on the next page will assist you in solving this type of problem. It is often most efficient to organize the data into a table.

MOMENTUM

Method for Solving Conservation of Momentum Problems

1. Identify the objects in the system. Confirm that the objects comprise an isolated system. Select a subscript for each object that describes the object, if possible. For example, use a subscript of *g* for gun, and *b* for bullet, rather than just 1 or 2.
2. Does the number of objects in the system change with the event? For example, if two objects stick together, after the collision there will be only one object with a mass equal to the total of the masses of the two individual objects.
3. Determine the momentum of each object in the system before the event. Use the proper variable for any unknown quantities.
4. Determine the momentum of each object in the system after the event. Likewise, use the proper variable for any unknowns.
5. Set the total in #3 above equal to the total in #4 above. Solve for the unknown variable.

Sample Example Problems

1. Cart A, of mass 1.0 kg, is at rest on a frictionless air track. It is struck by Cart B, of mass 0.20 kg, moving to the right with a speed of 10. m/s. After the collision, Cart A is moving to the right with a speed of 3.0 m/s. What is the velocity of Cart B after the collision?

$$p_{before} = p_{after}$$

$$m_A v_A + m_B v_B = m_A v'_A + m_B v'_B$$

$$(1.0 kg)(0) + (0.20 kg)(10. m/s) = (1.0 kg)(3.0 m/s) + (0.20 kg)v'_B$$

$$v'_B = -5.0 m/s$$

2. Calculate the recoil velocity of a 4.0 kg rifle that shoots a 0.050-kg bullet at a speed of 280 m/s.

$$p_{before} = p_{after}$$

$$(m_r + m_b)v = m_r v'_r + m_b v'_b$$

$$(4.0 kg + 0.050 kg)0 = (4.0 kg)v'_r + (0.050 kg)(280 m/s)$$

$$0 = (4.0 kg)v'_r + (0.050 kg)(280 m/s)$$

$$v'_r = -3.5 m/s$$

3. A 2.5×10^4 kilogram railroad car moving at 4.0 m/s east collides with another car of the same mass moving east at 1.5 m/s. The two cars couple together after the collision. Determine the final speed of the two cars after they have collided.

$$p_{before} = p_{after}$$

$$m_1 v_1 + m_2 v_2 = (m_1 + m_2)v'$$

$$(2.5 \times 10^4 kg)(4.0 m/s) + (2.5 \times 10^4 kg)(1.5 m/s) = (2.5 \times 10^4 kg + 2.5 \times 10^4 kg)v'$$

$$v' = 2.8 m/s$$

Unit 5
Practice Questions

1. Cart A has a mass of 2 kilograms and a speed of 3 meters per second. Cart B has a mass of 3 kilograms and a speed of 2 meters per second. Compared to the inertia and magnitude of momentum of cart A, cart B has

   (1) the same inertia and a smaller magnitude of momentum
   (2) the same inertia and the same magnitude of momentum
   (3) greater inertia and a smaller magnitude of momentum
   (4) greater inertia and the same magnitude of momentum

2. A 0.45-kilogram football traveling at a speed of 22 meters per second is caught by an 84-kilogram stationary receiver. If the football comes to rest in the receiver's arms, the magnitude of the impulse imparted to the receiver by the ball is

   (1) 1800 N•s   (2) 9.9 N•s   (3) 4.4 N•s   (4) 3.8 N•s

3. The diagram below represents two masses before and after they collide. Before the collision, mass m_A is moving to the right with speed v, and mass m_B is at rest. Upon collision, the two masses stick together.

Before Collision   **After Collision**

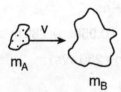

m_A m_B m_A m_B

Which expression represents the speed, v', of the masses after the collision? [Assume no outside forces are acting on mA or mB.]

(1) $\dfrac{m_A + m_B v}{m_A}$ (3) $\dfrac{m_B v}{m_A + m_B}$

(2) $\dfrac{m_A + m_B}{m_A v}$ (4) $\dfrac{m_A v}{m_A + m_B}$

4. A woman with horizontal velocity v_1 jumps off a dock into a stationary boat. After landing in the boat, the woman and the boat move with velocity v_2. Compared to velocity v_1, velocity v_2 has

(1) the same magnitude and the same direction
(2) the same magnitude and opposite direction
(3) smaller magnitude and the same direction
(4) larger magnitude and the same direction

5. A force of 6.0 newtons changes the momentum of a moving object by 3.0 kilogram•meters per second. How long did the force act on the mass?

(1) 1.0 s   (2) 2.0 s   (3) 0.25 s   (4) 0.50 s

6. A 3.0-kilogram steel block is at rest on a frictionless horizontal surface. A 1.0-kilogram lump of clay is propelled horizontally at 6.0 meters per second toward the block as shown in the diagram below.

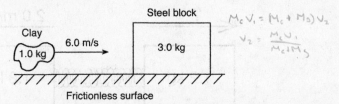

Upon collision, the clay and steel block stick together and move to the right with a speed of

(1) 1.5 m/s   (2) 2.0 m/s   (3) 3.0 m/s   (4) 6.0 m/s

7. A 2.0-kilogram body is initially traveling at a velocity of 40. meters per second east. If a constant force of 10. newtons due east is applied to the body for 5.0 seconds, the final speed of the body is

(1) 15 m/s   (2) 25 m/s   (3) 65 m/s   (4) 130 m/s

8. At the circus, a 100.-kilogram clown is fired at 15 meters per second from a 500.-kilogram cannon. What is the recoil speed of the cannon?

(1) 75 m/s   (2) 15 m/s   (3) 3.0 m/s   (4) 5.0 m/s

9. A 50.-kilogram student threw a 0.40-kilogram ball with a speed of 20. meters per second. What was the magnitude of the impulse that the student exerted on the ball?

(1) 8.0 N • s   (3) 4.0×10^2 N • s
(2) 78 N • s   (4) 1.0×10^3 N • s

10. What is the speed of a 1.0×10^3-kilogram car that has a momentum of 2.0×10^4 kilogram • meters per second east?

(1) 5.0×10^{-2} m/s

(2) 2.0×10^1 m/s

(3) 1.0×10^4 m/s

(4) 2.0×10^7 m/s

11. Ball A of mass 5.0 kilograms moving at 20. meters per second collides with ball B of unknown mass moving at 10. meters per second in the same direction. After the collision, ball A moves at 10. meters per second and ball B at 15 meters per second, both still in the same direction. What is the mass of ball B?

(1) 6.0 kg

(2) 2.0 kg

(3) 10. kg

(4) 12 kg

12. A 1.2-kilogram block and a 1.8-kilogram block are initially at rest on a frictionless, horizontal surface. When a compressed spring between the blocks is released, the 1.8-kilogram block moves to the right at 2.0 meters per second, as shown.

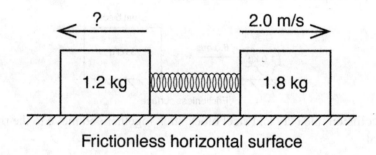

Frictionless horizontal surface

What is the speed of the 1.2-kilogram block after the spring is released?

(1) 1.4 m/s

(2) 2.0 m/s

(3) 3.0 m/s

(4) 3.6 m/s

13. Which is an acceptable unit for impulse?

(1) N•m

(2) J/s

(3) J•s

(4) kg•m/s

14. A 0.10-kilogram model rocket's engine is designed to deliver an impulse of 6.0 newton-seconds. If the rocket engine burns for 0.75 second, what average force does it produce?

(1) 4.5 N

(2) 8.0 N

(3) 45 N

(4) 80. N

15. The diagram below shows a 4.0-kilogram cart moving to the right and a 6.0-kilogram cart moving to the left on a horizontal frictionless surface.

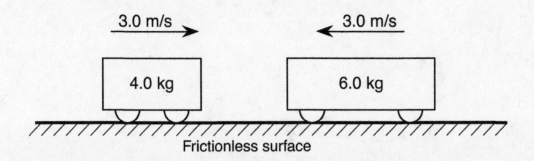

When the two carts collide they lock together. The magnitude of the total momentum of the two-cart system after the collision is

(1) 0.0 kg•m/s   (3) 15 kg•m/s

(2) 6.0 kg•m/s   (4) 30. kg•m/s

$(5)(20) + (10)M_B = (5)(10) + M_B(15)$

$50 = 5M_B$

$J = Ft$

$J = NS$

$J = \frac{15am}{5v} \cdot \frac{8}{1}$

$J = \frac{15am}{5}$

NOTES:

UNIT 6
Energy

One of the central definitions of physics is that it is a study of the interactions among matter and energy. Energy is often described as the ability to do work.

Work

In order for work to be done on an object, there must be a *force* exerted on that object resulting in a *displacement of that object*. If the applied force does not move the object, no work is done on that object. For example, pushing a crate across the floor results in work being done. However, holding a textbook out at arm's length, motionless, requires no work. There is a force, but no displacement. In equation form, work is calculated by:

$$W = Fd$$

The net work done on an object is equal to the change in the total energy of the object.

This important result is known as the the *Work-Energy Theorem*.

$$W = Fd = \Delta E_t$$

The equation above defines the unit of work as the Newton-meter (Nm). This is equivalent to a Joule (J), the SI unit for work, named after James Prescott Joule. One joule represents the amount of work done when an object is displaced one meter by a force of one Newton. Work is a scalar quantity.

In some resources, the equation for work is refined to include a subscript on the Force term:

$$W = F_\parallel d$$

This subscript denotes that the force is *parallel to* the displacement. While many forces may act on an object, the only force doing work is the <u>force that is in the same direction as the displacement</u>. Often, the component of a force that is parallel to the displacement must be calculated before determining the work done. For example, if a person were to pull a crate by a rope attached to it at an angle, only the component of the force parallel to the displacement does work on the crate, as in the diagram below.

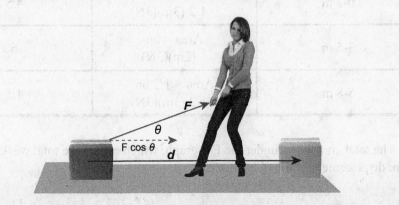

The component of the force parallel to the displacement can be determined from right triangle trigonometry as the product of the original force (the hypotenuse) and the cosine of the angle it makes with the horizontal. Therefore, the equation for work can now be expressed as:

$$W = Fd \cos\theta$$

Work Done by a Varying Force

In some situations, the force exerted on an object varies and is not constant. In order to determine the work done, a graph of force vs. displacement is plotted, as in the example below.

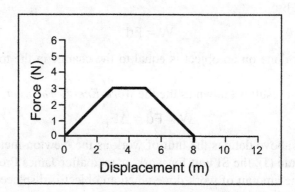

In order to determine the total work done, the *area under the curve* must be calculated from this graph. Recall from earlier units that the area under the curve is determined by breaking up the graph into regions of rectangles and triangles and determining the area of these resulting shapes, taking into account the scales used on each axis. For the graph above, the total work done for the entire displacement of 10 meters is:

Interval	Area calculation	Result
0-3 m	Area = 1/2 bh = 1/2 (3m)(3N)	4.5 J
3-5 m	Area = bh = (2m)(3N)	6 J
5-8 m	Area = 1/2 bh = 1/2 (3m)(3N)	4.5 J

The total area found under the F-d graph, which equals the total work done for the entire displacement of the object, is 15 J.

Power

Simply put, **power** is the rate that work gets done. Since power refers to a rate, it includes time in the equation:

$$P = \frac{W}{t}$$

This equation can be expanded to include the formula for work, and then further simplified in terms of average velocity:

$$P = \frac{Fd}{t} = F\bar{v}$$

Power is a scalar quantity and the unit is Joules/second (J/s), or the Watt (W), after Scottish engineer James Watt.

James Watt

James Watt was a Scottish mechanical engineer and inventor most noted for his work improving the steam engine. In addition to this work, he also invented and developed a machine to make exact copies of documents by transferring the ink directly to another piece of paper. The SI unit of power (the Watt, W) is named after him.

Born: January 19, 1736; Scotland
Died: August 25, 1819; England

A very popular lab activity is to determine a student's *horsepower*. The term horsepower, invented by James Watt, is often used to represent the power of things like car and motorcycle engines. One horsepower is equal to 746 Watts (1hp = 746W).

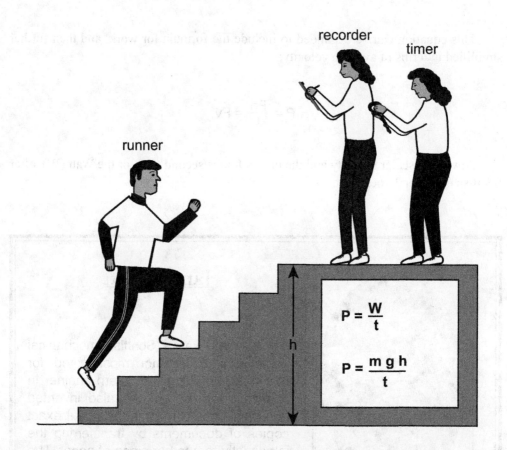

In the activity, students measure the total height (h) of a flight of stairs to be climbed. The force necessary to climb the stairs is equal to the student's weight in Newtons (m•g). If a student were to maximize their power, that student would want to climb the stairs in the smallest amount of time possible, due to the inverse relationship between power and time.

Using the equation above, the student's power in Watts can be calculated. It can then be converted into horsepower, if desired, to compare to other everyday objects and machines.

Solved Example Problem

- A horizontal force of 805 N is needed to drag a crate across a horizontal floor with a constant speed. Jason drags the crate using a rope held at an angle of 32°.

    (a) What force does Jason exert on the rope?

    $$F_\parallel = F \cos \theta$$

    $$805 \text{ N} = F \cos 32°$$

    $$F = 950 \text{ N}$$

    (b) How much work does Jason do on the crate when moving it 22 m?

    $$W = F_\parallel d$$

    $$W = (805 \text{ N})(22 \text{ m})$$

    $$W = 18000 \text{ J}$$

    (c) If Jason completes the job in 8.0 s, what power is developed?

    $$P = \frac{W}{t}$$

    $$P = \frac{18000 \text{ J}}{8.0 \text{ s}}$$

    $$P = 2300 \text{ W}$$

Energy

As the definition of work states, the amount of work done will change the total energy of an object. So what is energy? One definition of energy is that it is the *ability to do work*. Since work changes an object's energy, and if an object has energy it has the capability to do work, the unit of energy must then be the same as the unit for work. This is the nature of the *Work-energy Theorem*. Both quantities are measured with the Joule (J).

Types of Energy

Generally speaking, there are three main types of energy: **potential energy, kinetic energy, and internal energy**. These three types of energy combined represent the *total mechanical energy* of an object or system of objects.

$$E_T = PE + KE + Q$$

where:
E_T = total mechanical energy
PE = potential energy
KE = kinetic Energy
Q = internal energy

Potential Energy

Potential energy is often referred to as stored energy. An object with potential energy has the capability to convert this energy into other forms, such as kinetic energy. This stored energy is due to the object's position or condition. There are two main types of potential energy: *gravitational potential energy* and *elastic potential energy.* Other types include chemical, nuclear, magnetic, and electrical potential energies.

Gravitational Potential Energy is the stored energy due to an object's relative position in a gravitational field. Since gravitational field strength varies with distance from the earth's center (height), then the gravitational potential energy varies with height as well.

Suppose that a weight lifter were to lift a barbell from the ground to a particular height, (h). In order to lift this barbell, the weight (m•g) must be lifted a particular height (h). This lifting certainly requires work to be done. Recall that work is the product of force and displacement. In this case, the force required to lift the barbell is equal to its weight, and the displacement is equal to the height it is lifted above the earth's surface.

In this example, a height of zero is defined as the earth's surface. In equation form, the work done to lift the barbell is:

$$W = Fd = mgh = \Delta E_T$$

The work done in lifting the barbell is equal to the change in energy of the barbell. At the top of the lift, the barbell has more energy than on the earth's surface. Since the barbell is not moving at the top, it does not have kinetic energy, and the type of energy it has gained is gravitational potential energy. The work done is equal to the change in gravitational potential energy. Therefore, the change in gravitational potential energy is:

$$\Delta PE = mg\Delta h$$

This equation states that the amount of gravitational potential energy is directly related to that object's height above a certain reference point. A height of zero is typically defined as the lowest point the object could possibly travel to (often the earth's surface). There are some situations where a height of zero is defined differently, such as situations where the lowest point might be a tabletop, but this is not as common as using the earth's surface as a height of zero.

Other objects can store energy based on their ability to stretch, or deform their shapes, and then return to their original shapes. These objects are said to be *elastic*. Examples of elastic objects include springs, rubber bands, and metal strips. When these objects are stretched, they gain **elastic potential energy**. Again, considering the *work-energy theorem*, stretching the object requires work to be done. For example, in the sport of archery (bow and arrow), work must be done in order to pull the arrow against the bowstring.

As before, the work done is equal to the product of the force and the displacement, which also equals the change in elastic potential energy when stretched.

$$W = Fd = \Delta E_t$$

In stretching an object, the force is not constant. The farther the object is stretched, the greater the force necessary to stretch it. This relationship is called **Hooke's Law**, named after British physicist Robert Hooke.

Robert Hooke

Robert Hooke was an English philosopher, architect, and mathematician. Although often overshadowed by Newton, many consider Hooke one of the greatest experimental scientists of the 17th century. His experiments included development of the universal joint, invention of the watch balance spring, and development of an equation describing elasticity (Hooke's Law). He was also an early proponent of evolution.

Born: July 18, 1635; England
Died: March 3, 1703 ; England

Algebraically, Hooke's Law is stated as:

$$F = kx$$

where:
F = force
k = spring constant
x = elongation of the spring from the equilibrium position

The spring constant is a property of a particular elastic object. It is a constant for that object, but each object may have a different spring constant. If the equation above were solved for k, it would become a ratio of F/x. This ratio is the same for any displacement of a given spring, as long as the elastic limit is not exceeded, which damages the spring. In the diagram below, the forces necessary for a variety of stretches of the same spring are shown.

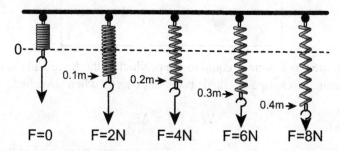

The spring constant can be calculated from any of the parts of the diagram above. Example: for a stretch of 0.3 m from the equilibrium position:

$$k = F/x$$
$$k = 6N/0.3m$$
$$k = 20 \text{ N/m}$$

Another way to determine the spring constant is to plot the force necessary to stretch the spring vs. the elongation of the spring on a graph. The *slope* of the graph is the spring constant, k.

slope = $k = \Delta F / \Delta x$

$k = (8N - 0)/(0.4m - 0)$

$k = 20$ N/m

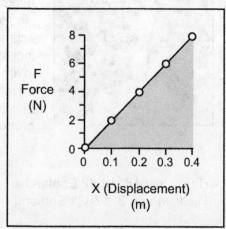

The work done in stretching the spring is equal to the gain in elastic potential energy of the spring. There are two ways to determine this. One way is to determine the *area under the curve* of the graph. The shape formed in the graph (the shaded region) is a triangle. The formula for the area of a triangle is $\frac{1}{2}$bh. The base is the elongation (x), and the height is the force (F). Replacing the force with kx from Hooke's Law, the area under the curve becomes:

$$\text{Area} = \frac{1}{2}bh$$
$$= \frac{1}{2}(x)(F)$$
$$= \frac{1}{2}(x)(kx)$$
$$\text{Area} = PE_s = \frac{1}{2}kx^2$$

The other way to determine the elastic potential energy is to go back to the equation for work, Fd. In this equation, the force represents the *average force*. In many other situations, the force is constant so this was not really an issue. For an elastic object, however, the force increases linearly with elongation. The average force can be determined if the initial force (zero) and maximum force are known:

$$F_{avg} = 0 + \frac{F_{max}}{2}$$

Leaving off the subscript of *max* and replacing the *d* with *x*, the work formula for a spring can be expressed as:

$$W = \frac{1}{2}Fx$$

From Hooke's Law, *kx* can be substituted for *F* above. This is the work done (and therefore the elastic potential energy gained) in stretching a spring.

$$PE_s = \frac{1}{2}kx^2$$

Kinetic Energy

The energy of motion is called **Kinetic Energy** (KE). Kinetic energy is a scalar quantity and the unit of KE is the Joule (J), the same unit as work and the other forms of energy. The formula for kinetic energy can be derived from other known equations for work, net force, and kinematics:

$$W = Fd \ \ (\text{using } F_{net} = ma, \text{ substitute } ma \text{ for } F)$$

$$W = ma \cdot d \ \ (\text{using } v_f^2 = v_i^2 + 2ad \text{ where } v_i = 0, \text{ substitute } \frac{v^2}{2} \text{ for } ad)$$

$$W = \frac{mv^2}{2}$$

If all of the work done is to increase the speed of an object from rest (or stop a moving object), this can be the equation for kinetic energy:

$$KE = \frac{1}{2}mv^2$$

Conservation of Energy

Unless there is work done by an agent outside a system, the total energy of that object or system of objects will not change. The energy, however, can transform from one type to another, or transfer from one object to another in a system. This is called the **Law of Conservation of Energy**.

Recall that the total energy of an object or system is the sum of the kinetic, potential, and internal energies.

$$E_T = KE + PE + Q$$

There are several excellent examples that show the law of conservation of energy in action. The two most notable are the actions of a pendulum and that of a simple roller coaster.

In a pendulum, work must initially be done in order to pull back the pendulum bob (mass) so that it can swing freely. As a pendulum swings, it demonstrates *simple harmonic* motion. This means that the motion continually repeats itself. Without friction, an ideal pendulum will continue to swing forever with a constant amplitude (height of swing), and period (time to swing back and forth). It should be noted that the only factor to affect the period of a pendulum is the length of the string from which it swings. The period of a pendulum can be calculated as follows:

$$T = 2\pi\sqrt{\frac{l}{g}}$$

where:
T = period (measured in seconds)
l = length of string
g = acceleration due to gravity (9.81 m/s^2 near the surface of earth)

According to the work-energy theorem, at this point the total energy of the pendulum has increased. If there are no other external forces, the total energy of the pendulum will remain constant upon release.

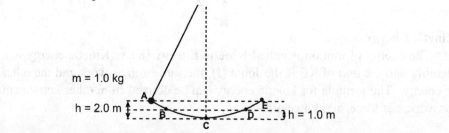

At the moment of release (A in the diagram above), the pendulum has no kinetic energy and all gravitational potential energy. The amount of energy it has is equal to the work done in raising it to this height. As the pendulum falls toward the lowest point (C), the kinetic energy is increasing and the gravitational potential energy is decreasing. While falling (B) or rising (D), the pendulum has a combination of some kinetic energy and some gravitational potential energy. Defining point (C) as the lowest point and at a height (h) of zero, there is no gravitational potential energy and the total energy is now all kinetic energy. As the pendulum passes through the lowest point and heads toward the other endpoint (E), the kinetic energy is decreasing and the gravitational potential energy is increasing. When the pendulum reaches the other end (E), it has risen to the same height from which it was initially released and at this point the total energy is all gravitational potential energy. This process continues as the pendulum continues to swing. Throughout the entire time, the total energy is constant. This example is solved numerically in the table below.

LOCATION	PE (mgh)	KE ($\frac{1}{2}$ mv^2)	E$_T$ (PE+KE)	v (2 KE/m)$^{\frac{1}{2}}$
A or E	1.0 kg × 9.8 m/s^2 ™ 2.0 m 19.6 J	0	19.6 J	0
C	0	19.6 J	19.6 J	6.26 m/s
B or D	1.0 kg × 9.8 m/s^2 × 1.0 m 9.8 J	9.8 J	19.6 J	4.43 m/s

In a roller coaster, work is done to pull the coaster train up to the top of the first hill. This process gives the train all the mechanical energy it needs to travel along the track. Unless there is an outside force (another lift hill or a braking section), the total energy of the roller coaster train is constant from this point forward. At the top of the first hill, the train has maximum gravitational potential energy and very little (or none if motionless) kinetic energy. As the train goes down the first hill, the gravitational potential energy decreases and the kinetic energy increases. At the bottom of the first hill, if it is the lowest point on the ride, the train has all kinetic energy and no gravitational potential energy. As it travels up the second hill, the train gains gravitational potential energy and loses kinetic energy. This process continues as it travels up and down subsequent hills.

You may notice that the first hill on most roller coasters is the highest; the other hills are all lower. This is because in any real system such as a roller coaster or a pendulum, there is always **friction**, which *appears* to cause the total energy to decrease. Since total mechanical energy is conserved and can not change, where does this energy go? Since friction is a force, it does work on the object or system of objects. This work causes the total kinetic and potential energies to decrease, but it also causes the *internal energy* to increase, keeping the total energy (E$_T$) constant. Therefore, the Law of Conservation of Energy is valid.

Solved Example - Conservation of Energy

- A 4.00 kg block oscillates vertically on the end of a spring in simple harmonic motion. If the elastic potential energy is zero at its highest point, find the following for the system's lowest point, midpoint, and highest point in its motion. Assume that h = 0 m at the lowest point, and h = 0.600 m at the highest point.
  - elastic potential energy (PE_s)
  - kinetic energy (KE)
  - gravitational potential energy (PE_g)
  - total energy (E_T)

	Elastic Potential Energy $PE_s = \frac{1}{2}kx^2$	Kinetic Energy $KE = \frac{1}{2}mv^2$	Gravitational Potential Energy $PE_g = mgh$	Total Energy E_T
Lowest Point	$23.5\ J = \frac{1}{2}k(0.600m)^2$ $k = 131\ kg/s^2$ **$PE_s = 23.5\ J$** **(Due to Conservation of Energy)**	$KE = \frac{1}{2}mv^2$ (v = 0) **∴ KE = 0**	$PE_g = mgh$ (h = 0) **∴ PEg = 0**	23.5 J
Equilibrium Position	$PE_s = \frac{1}{2}kx^2$ $PE_s = \frac{1}{2}(131 kg/s^2)(0.300m)^2$ **$PE_s = 5.90\ J$**	$KE = \frac{1}{2}mv^2$ $5.80\ J = \frac{1}{2}(4.00\ kg)v^2$ V = 1.70 m/s **KE = 5.80 J** **(Due to Conservation of Energy)**	$PE_g = mgh$ $PE_g = (4.00kg)(9.81m/s^2)(0.300m)$ **PEg = 11.8 J**	23.5 J
Highest Point	$PE_s = \frac{1}{2}kx^2$ (x = 0) **∴ PEs = 0**	$KE = \frac{1}{2}mv^2$ (v = 0) **∴ KE = 0**	$PE_g = mgh$ $PE_g = (4.00kg)(9.81m/s^2)(0.600m)$ **PEg = 23.5 J**	23.5 J

Unit 6
Practice Questions

1. What is the average power required to raise a 1.81×10^4-newton elevator 12.0 meters in 22.5 seconds?

   (1) 8.04×10^2 W   (3) 2.17×10^5 W

   (2) 9.65×10^3 W   (4) 4.89×10^6 W

2. If the speed of a moving object is doubled, the kinetic energy of the object is

   (1) halved   (2) doubled   (3) unchanged   (4) quadrupled

3. How much work is required to lift a 10.-newton weight from 4.0 meters to 40. meters above the surface of Earth?

   (1) 2.5 J   (2) 3.6 J   (3) 3.6×10^2 J   (4) 4.0×10^2 J

4. A child does 0.20 joule of work to compress the spring in a pop-up toy. If the mass of the toy is 0.010 kilogram, what is the maximum vertical height that the toy can reach after the spring is released?

   (1) 20. m   (2) 2.0 m   (3) 0.20 m   (4) 0.020 m

5. A book of mass m falls freely from rest to the floor from the top of a desk of height h. What is the speed of the book upon striking the floor?

   (1) $\sqrt{2gh}$ (2) 2gh   (3) mgh   (4) mh

6. A car travels at constant speed v up a hill from point A to point B, as shown in the diagram below.

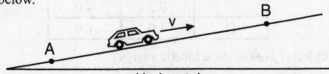

   As the car travels from A to B, its gravitational potential energy

   (1) increases and its kinetic energy decreases

   (2) increases and its kinetic energy remains the same

   (3) remains the same and its kinetic energy decreases

   (4) remains the same and its kinetic energy remains the same

7. Which combination of fundamental units can be used to express energy?

$W = Fd$
$W = \frac{kg \cdot m}{s^2} \cdot m$

(1) kg • m/s   (2) kg • m²/s   (3) kg • m/s²   (4) kg • m²/s²

8. An object is thrown vertically upward. Which pair of graphs best represents the object's kinetic energy and gravitational potential energy as functions of its displacement while it rises?

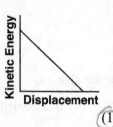
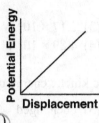

(1)

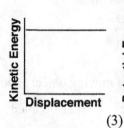
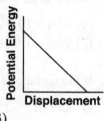

(3)

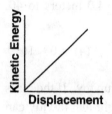
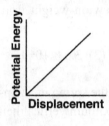

(2)

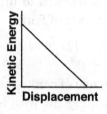
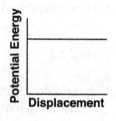

(4)

9. The table below lists the mass and speed of each of four objects.

Objects	Mass (kg)	Speed (m/s)
A	1.0	4.0
B	2.0	2.0
C	0.5	4.0
D	4.0	1.0

Which two objects have the same kinetic energy?

(1) A and D   (2) B and D   (3) A and C   (4) B and C

10. A spring with a spring constant of 80. newtons per meter is displaced 0.30 meter from its equilibrium position. The potential energy stored in the spring is

(1) 3.6 J   (2) 7.2 J   (3) 12 J   (4) 24 J

11. A pendulum is pulled to the side and released from rest. Which graph best represents the relationship between the gravitational potential energy of the pendulum and its displacement from its point of release?

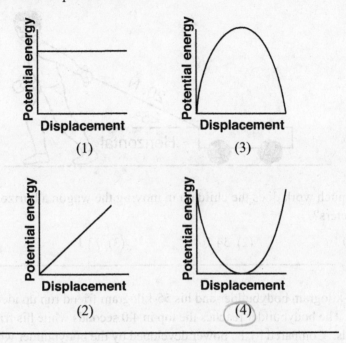

(1)

(3)

(2)

(4)

12. The work done in accelerating an object along a frictionless horizontal surface is equal to the change in the object's

(1) momentum   (3) potential energy

(2) velocity   (4) kinetic energy

13. As a block slides across a table, its speed decreases while its temperature increases. Which two changes occur in the block's energy as it slides?
(1) a decrease in kinetic energy and an increase in internal energy
(2) an increase in kinetic energy and a decrease in internal energy
(3) a decrease in both kinetic energy and internal energy
(4) an increase in both kinetic energy and internal energy

14. A 1.00-kilogram ball is dropped from the top of a building. Just before striking the ground, the ball's speed is 12.0 meters per second. What was the ball's gravitational potential energy, relative to the ground, at the instant it was dropped? [Neglect friction]

(1) 6.00 J   (2) 24.0 J   (3) 72.0 J   (4) 144 J

15. As shown in the diagram below, a child applies a constant 20.-newton force along the handle of a wagon which makes a 25° angle with the horizontal.

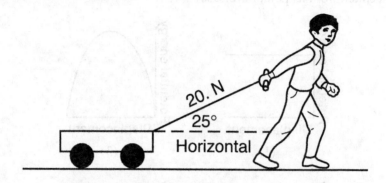

How much work does the child do in moving the wagon a horizontal distance of 4.0 meters?

(1) 5.0 J   (2) 34 J   (3) 73 J   (4) 80. J

---

16. A 110-kilogram bodybuilder and his 55-kilogram friend run up identical flights of stairs. The bodybuilder reaches the top in 4.0 seconds while his friend takes 2.0 seconds. Compared to the power developed by the bodybuilder while running up the stairs, the power developed by his friend is

(1) the same   (3) half as much
(2) twice as much   (4) four times as much

17. A 2.0-kilogram block sliding down a ramp from a height of 3.0 meters above the ground reaches the ground with a kinetic energy of 50. joules. The total work done by friction on the block as it slides down the ramp is approximately

(1) 6 J   (2) 9 J   (3) 18 J   (4) 44 J

18. During an emergency stop, a 1.5×10^3-kilogram car lost a total of 3.0×10^5 joules of kinetic energy. What was the speed of the car at the moment the brakes were applied?

(1) 10. m/s   (2) 14 m/s   (3) 20. m/s   (4) 25 m/s

19. A 6.8-kilogram block is sliding down a horizontal, frictionless surface at a constant speed of 6.0 meters per second. The kinetic energy of the block is approximately

(1) 20. J   (2) 41 J   (3) 120 J   (4) 240 J

ENERGY

20. Through what vertical distance is a 50.-newton object moved if 250 joules of work is done against the gravitational field of Earth?

(1) 2.5 m (2) 5.0 m (3) 9.8 m (4) 25 m

21. A 55.0-kilogram diver falls freely from a diving platform that is 3.00 meters above the surface of the water in a pool. When she is 1.00 meter above the water, what are her gravitational potential energy and kinetic energy with respect to the water's surface?

(1) PE = 1620 J and KE = 0 J (3) PE = 810 J and KE = 810 J
(2) PE = 1080 J and KE = 540 J (4) PE = 540 J and KE = 1080 J

22. A truck weighing 3.0×10^4 newtons was driven up a hill that is 1.6×10^3 meters long to a level area that is 8.0×10^2 meters above the starting point. If the trip took 480 seconds, what was the minimum power required?

(1) 5.0×10^4 W (3) 1.2×10^{10} W
(2) 1.0×10^5 W (4) 2.3×10^{10} W

23. As shown in the diagram below, a student exerts an average force of 600. newtons on a rope to lift a 50.0-kilogram crate a vertical distance of 3.00 meters.

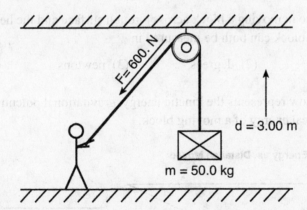

Compared to the work done by the student, the gravitational potential energy gained by the crate is

(1) exactly the same (3) 330 J more
(2) 330 J less (4) 150 J more

---

24. A 60.0-kilogram runner has 1920 joules of kinetic energy. At what speed is she running?

(1) 5.66 m/s (2) 8.00 m/s (3) 32.0 m/s (4) 64.0 m/s

25. The diagram below shows points A, B, and C at or near Earth's surface. As a mass is moved from A to B, 100. joules of work are done against gravity.

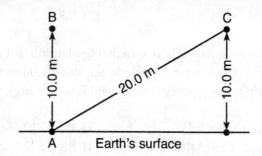

What is the amount of work done against gravity as an identical mass is moved from A to C ?

(1) 100. J   (2) 173 J   (3) 200. J   (4) 273 J

---

26. A motor used 120. watts of power to raise a 15-newton object in 5.0 seconds. Through what vertical distance was the object raised?

(1) 1.6 m   (2) 8.0 m   (3) 40. m   (4) 360 m

27. The work done in moving a block across a rough surface and the heat energy gained by the block can both be measured in

(1) watts   (2) degrees   (3) newtons   (4) joules

28. The graph below represents the kinetic energy, gravitational potential energy, and total mechanical energy of a moving block.

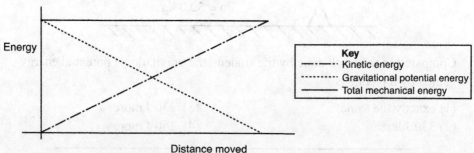

Which best describes the motion of the block?

(1) accelerating on a flat horizontal surface   (3) falling freely
(2) sliding up a frictionless incline   (4) being lifted at constant velocity

29. The diagram below shows a 50.-kilogram crate on a frictionless plane at angle θ to the horizontal. The crate is pushed at constant speed up the incline from point A to point B by force F.

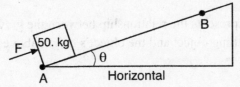

If angle θ were increased, what would be the effect on the magnitude of force F and the total work W done on the crate as it is moved from A to B?

(1) W would remain the same and the magnitude of F would decrease.
(2) W would remain the same and the magnitude of F would increase.
(3) W would increase and the magnitude of F would decrease.
(4) W would increase and the magnitude of F would increase.

30. As a ball falls freely (without friction) toward the ground, its total mechanical energy

(1) decreases   (2) increases   (3) remains the same

31. A 0.50-kilogram ball is thrown vertically upward with an initial kinetic energy of 25 joules. Approximately how high will the ball rise? [Neglect air resistance.]

(1) 2.6 m   (2) 5.1 m   (3) 13 m   (4) 25 m

32. The amount of work done against friction to slide a box in a straight line across a uniform, horizontal floor depends most on the

(1) time taken to move the box   (3) speed of the box
(2) distance the box is moved   (4) direction of the box's motion

33. A block weighing 15 newtons is pulled to the top of an incline that is 0.20 meter above the ground, as shown below.

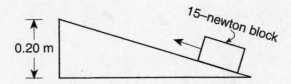

If 4.0 joules of work are needed to pull the block the full length of the incline, how much work is done against friction?

(1) 1.0 J   (2) 0.0 J   (3) 3.0 J   (4) 7.0 J

34. In raising an object vertically at a constant speed of 2.0 meters per second, 10. watts of power are developed. The weight of the object is

(1) 5.0 N   (2) 20. N   (3) 40. N   (4) 50. N

35. Which graph best represents the relationship between the gravitational potential energy of a freely falling object and the object's height above the ground near the surface of Earth?

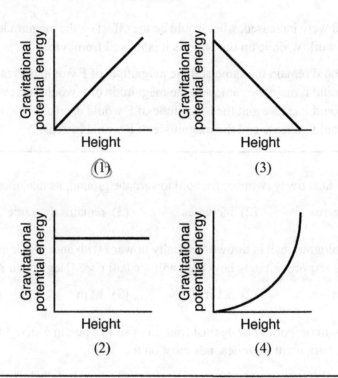

UNITS 1-6 Cumulative Review
Part A and B1

1. What is the approximate length of a baseball bat?

   (1) 10^{-1} m   (3) 10^1 m
   (2) 10^0 m   (4) 10^2 m

2. A force of 1 newton is equivalent to

   (1) $\dfrac{kg \cdot m}{s^2}$ (2) $\dfrac{kg \cdot m}{s}$ (3) $\dfrac{kg \cdot m^2}{s^2}$ (4) $\dfrac{kg^2 \cdot m^2}{s^2}$

3. A 6.0-newton force and an 8.0-newton force act concurrently on a point. As the angle between these forces increases from 0° to 90°, the magnitude of their resultant

   (1) decreases   (2) increases   (3) remains the same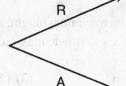

4. If the direction of a moving car changes and its speed remains constant, which quantity must remain the same?
   (1) velocity   (3) displacement
   (2) momentum   (4) kinetic energy

5. Forces A and B have a resultant R. Force A and resultant R are represented in the diagram below.

   R

   A

   Which vector best represents force B?

   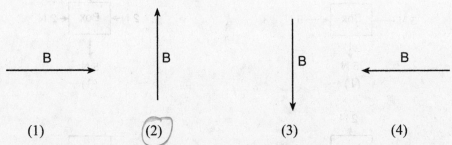

   B   B   B   B

   (1)   (2)   (3)   (4)

6. An astronaut standing on a platform on the Moon drops a hammer. If the hammer falls 6.0 meters vertically in 2.7 seconds, what is its acceleration?

   (1) 1.6 m/s^2 (2) 2.2 m/s^2 (3) 4.4 m/s^2 (4) 9.8 m/s^2

7. The speed of a wagon increases from 2.5 meters per second to 9.0 meters per second in 3.0 seconds as it accelerates uniformly down a hill. What is the magnitude of the acceleration of the wagon during this 3.0-second interval?

(1) 0.83 m/s²　　　(2) 2.2 m/s²　　　(3) 3.0 m/s²　　　(4) 3.8 m/s²

8. The graph below represents the relationship between speed and time for an object moving along a straight line.

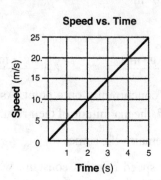

What is the total distance traveled by the object during the first 4 seconds?

(1) 5 m
(2) 20 m
(3) 40 m
(4) 80 m

9. A car initially traveling at a speed of 16 meters per second accelerates uniformly to a speed of 20. meters per second over a distance of 36 meters. What is the magnitude of the car's acceleration?

(1) 0.11 m/s²　　　(2) 2.0 m/s²　　　(3) 0.22 m/s²　　　(4) 9.0 m/s²

10. After a model rocket reached its maximum height, it then took 5.0 seconds to return to the launch site. What is the approximate maximum height reached by the rocket? [Neglect air resistance.]

(1) 49 m　　　(2) 98 m　　　(3) 120 m　　　(4) 250 m

11. Which diagram represents a box in equilibrium?

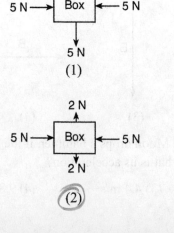

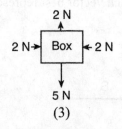

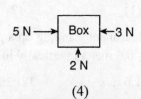

12. A 25-newton horizontal force northward and a 35-newton horizontal force southward act concurrently on a 15-kilogram object on a frictionless surface. What is the magnitude of the object's acceleration?

    (1) 0.67 m/s^2 (2) 1.7 m/s^2 (3) 2.3 m/s^2 (4) 4.0 m/s^2

13. Which object has the greatest inertia?

    (1) a 5.0-kg object moving at a speed of 5.0 m/s
    (2) a 10.-kg object moving at a speed of 3.0 m/s
    (3) a 15-kg object moving at a speed of 1.0 m/s
    (4) a 20.-kg object at rest

14. A satellite weighs 200 newtons on the surface of Earth. What is its weight at a distance of one Earth radius above the surface of Earth?

    (1) 50 N   (2) 100 N   (3) 400 N   (4) 800 N

15. A man standing on a scale in an elevator notices that the scale reads 30 newtons greater than his normal weight. Which type of movement of the elevator could cause this greater-than-normal reading?

    (1) accelerating upward
    (2) accelerating downward
    (3) moving upward at constant speed
    (4) moving downward at constant speed

16. If the sum of all the forces acting on a moving object is zero, the object will

    (1) slow down and stop
    (2) change the direction of its motion
    (3) accelerate uniformly
    (4) continue moving with constant velocity

17. A net force of 25 newtons is applied horizontally to a 10.-kilogram block resting on a table. What is the magnitude of the acceleration of the block?

    (1) 0.0 m/s^2 (2) 0.26 m/s^2 (3) 0.40 m/s^2 (4) 2.5 m/s^2

18. An apple weighing 1 newton on the surface of Earth has a mass of approximately
    (1) 1×10^{-1} kg   (2) 1×10^{0} kg   (3) 1×10^{1} kg   (4) 1×10^{2} kg

19. A car moves with a constant speed in a clockwise direction around a circular path of radius r, as represented in the diagram below.

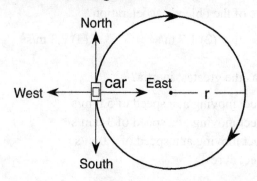

When the car is in the position shown, its acceleration is directed toward the

(1) north   (2) west   (3) south   (4) east

20. A ball attached to a string is moved at constant speed in a horizontal circular path. A target is located near the path of the ball as shown in the diagram.

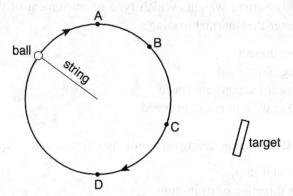

At which point along the ball's path should the string be released, if the ball is to hit the target?

(1) A   (2) B   (3) C   (4) D

21. A projectile is fired with an initial velocity of 120. meters per second at an angle, θ, above the horizontal. If the projectile's initial horizontal speed is 55 meters per second, then angle θ measures approximately

(1) 13°   (2) 27°   (3) 63°   (4) 75°

22. A 0.2-kilogram red ball is thrown horizontally at a speed of 4 meters per second from a height of 3 meters. A 0.4-kilogram green ball is thrown horizontally from the same height at a speed of 8 meters per second. Compared to the time it takes the red ball to reach the ground, the time it takes the green ball to reach the ground is

(1) one-half as great   (3) the same
(2) twice as great   (4) four times as great

Base your answers to questions 23 and 24 on the information below.

Projectile *A* is launched horizontally at a speed of 20. meters per second from the top of a cliff and strikes a level surface below, 3.0 seconds later. Projectile *B* is launched horizontally from the same location at a speed of 30. meters per second.

23. The time it takes projectile B to reach the level surface is

(1) 4.5 s   (2) 2.0 s   (3) 3.0 s   (4) 10. s

24. Approximately how high is the cliff?

(1) 29 m   (2) 44 m   (3) 60. m   (4) 104 m

25. Which situation will produce the greatest change of momentum for a 1.0-kilogram cart?

(1) accelerating it from rest to 3.0 m/s
(2) accelerating it from 2.0 m/s to 4.0 m/s
(3) applying a net force of 5.0 N for 2.0 s
(4) applying a net force of 10.0 N for 0.5 s

26. A 60-kilogram student jumps down from a laboratory counter. At the instant he lands on the floor his speed is 3 meters per second. If the student stops in 0.2 second, what is the average force of the floor on the student?

(1) 1×10^{-2}N   (2) 1×10^2N   (3) 9×10^2 N   (4) 4 N

27. In the diagram below, a block of mass M initially at rest on a frictionless horizontal surface is struck by a bullet of mass m moving with horizontal velocity v.

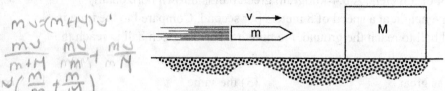

What is the velocity of the bullet-block system after the bullet embeds itself in the block?

(1) $\left(\frac{M+v}{M}\right)m$

(3) $\left(\frac{m+v}{M}\right)m$

(2) $\left(\frac{m+M}{m}\right)v$

(4) $\left(\frac{m}{m+M}\right)v$

28. During a collision, an 84-kilogram driver of a car moving at 24 meters per second is brought to rest by an inflating air bag in 1.2 seconds. The magnitude of the force exerted on the driver by the air bag is approximately

(1) 7.0×10^1N 　　(2) 8.2×10^2N 　　(3) 1.7×10^3 N 　　(4) 2.0×10^3 N

Base your answers to questions 29 and 30 on the information and diagram below.

The diagram shows a compressed spring between two carts initially at rest on a horizontal frictionless surface. Cart A has a mass of 2 kilograms and cart B has a mass of 1 kilogram. A string holds the carts together.

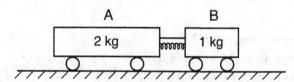

29. What occurs when the string is cut and the carts move apart?
(1) The magnitude of the acceleration of cart A is one-half the magnitude of the acceleration of cart B.
(2) The length of time that the force acts on cart A is twice the length of time the force acts on cart B.
(3) The magnitude of the force exerted on cart A is one-half the magnitude of the force exerted on cart B.
(4) The magnitude of the impulse applied to cart A is twice the magnitude of the impulse applied to cart B.
30. After the string is cut and the two carts move apart, the magnitude of which quantity is the same for both carts?
(1) momentum 　　(2) velocity 　　(3) inertia 　　(4) kinetic energy

31. An unstretched spring has a length of 10. centimeters. When the spring is stretched by a force of 16 newtons, its length is increased to 18 centimeters. What is the spring constant of this spring?

(1) 0.89 N/cm   (2) 2.0 N/cm   (3) 1.6 N/cm   (4) 1.8 N/cm

32. Which situation describes a system with *decreasing* gravitational potential energy?

(1) a girl stretching a horizontal spring   (3) a rocket rising vertically from Earth
(2) a bicyclist riding up a steep hill   (4) a boy jumping down from a tree limb

33. A 60.-kilogram student climbs a ladder a vertical distance of 4.0 meters in 8.0 seconds. Approximately how much total work is done against gravity by the student during the climb?

(1) 2.4×10^3 J   (3) 2.4×10^2 J
(2) 2.9×10^2 J   (4) 3.0×10^1 J

34. What is the maximum amount of work that a 6000.-watt motor can do in 10. seconds?

(1) 6.0×10^1 J   (3) 6.0×10^3 J
(2) 6.0×10^2 J   (4) 6.0×10^4 J

35. A horizontal force of 5.0 newtons acts on a 3.0-kilogram mass over a distance of 6.0 meters along a horizontal, frictionless surface. What is the change in kinetic energy of the mass during its movement over the 6.0-meter distance?

(1) 6.0 J   (2) 15 J   (3) 30. J   (4) 90. J

36. The potential energy stored in a compressed spring is to the change in the spring's length as the kinetic energy of a moving body is to the body's

(1) speed   (3) radius
(2) mass   (4) acceleration

37. An object falls freely near Earth's surface. Which graph best represents the relationship between the object's kinetic energy and its time of fall?

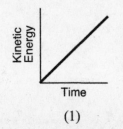   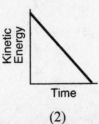   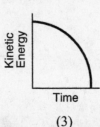   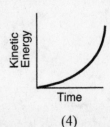

    (1)   (2)   (3)   (4)

38. A 95-kilogram student climbs 4.0 meters up a rope in 3.0 seconds. What is the power output of the student?

(1) 1.3×10^2 W

(2) 3.8×10^2 W

(3) 1.2×10^3 W

(4) 3.7×10^3 W

39. When a force moves an object over a rough, horizontal surface at a constant velocity, the work done against friction produces an increase in the object's

(1) weight

(2) momentum

(3) potential energy

(4) internal energy

40. A 1-kilogram rock is dropped from a cliff 90 meters high. After falling 20 meters, the kinetic energy of the rock is approximately

(1) 20 J

(2) 200 J

(3) 700 J

(4) 900 J

Base your answers to questions 1 and 2 on the information below:

The instant before a batter hits a 0.14-kilogram baseball, the velocity of the ball is 45 meters per second west. The instant after the batter hits the ball, the ball's velocity is 35 meter per second east. The bat and ball are in contact for 1.0 $\times 10^{-2}$ second.

1. Determine the magnitude and direction of the average acceleration of the baseball while it is in contact with the bat.

    Magnitude: _____ m/s^2

    Direction: _____

2. Calculate the magnitude of the average force the bat exerts on the ball while they are in contact. [Show all work, including the equation and substitution with units.]

_____

3. The graph below represents the velocity of an object traveling in a straight line as a function of time.

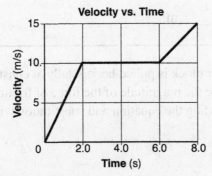

Velocity vs. Time

Determine the magnitude of the total displacement of the object at the end of the first 6.0 seconds.

_____ m

Base your answers to questions 4 through 6 on the information and vector diagram below.

A dog walks 8.0 meters due north and then 6.0 meters due east.

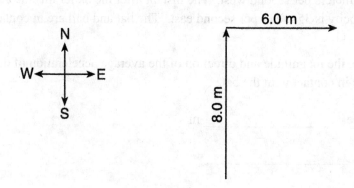

4. Using a metric ruler and the vector diagram, determine the scale used in the diagram.

    1.0 cm = _____ m

5. On the diagram above, construct the resultant vector that represents the dog's total displacement.

6. Determine the magnitude of the dog's total displacement.

    _____ m

    _____

7. A 10.-kilogram rubber block is pulled horizontally at constant velocity across a sheet of ice. Calculate the magnitude of the force of friction acting on the block. (Show all work, including the equation and substitution with units.)

Base your answer to questions 8 and 9 on the information below.

A car traveling at a speed of 13 meters per second accelerates uniformly to a speed of 25 meters per second in 5.0 seconds.

8. Calculate the magnitude of the acceleration of the car during this 5.0-second time interval. (Show all work, including the equation and substitution with units.)

9. A truck traveling at a constant speed covers the same total distance as the car in the same 5.0-second time interval. Determine the speed of the truck.

_____m/s

10. The gravitational force of attraction between Earth and the Sun is 3.52×10^{22} newtons. Calculate the mass of the Sun. (Show all work, including the equation and substitution with units.)

Base your answers to questions 11 and 12 on the information and graph below.

The graph represents the relationship between the force applied to each of two springs, *A* and *B*, and their elongations.

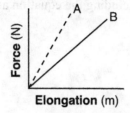

Force vs. Elongation

11. What physical quantity is represented by the slope of each line?

_____

12. A 1.0-kilogram mass is suspended from each spring. If each mass is at rest, how does the potential energy stored in spring *A* compare to the potential energy stored in spring *B*?

_____

_____

Base your answers to questions 13 and 14 on the information below.

The combined mass of a race car and its driver is 600. kilograms. Traveling at constant speed, the car completes one lap around a circular track of radius 160 meters in 36 seconds.

13. Calculate the speed of the car. (Show all work, including the equation and subsitution with units.)

14. Calculate the magnitude of the centripetal acceleration of the car. (Show all work, including the equation and substitution with units.)

---

15. An 8.00-kilogram ball is fired horizontally from a 1.00×10^3-kilogram cannon initially at rest. After having been fired, the momentum of the ball is 2.40×10^3 kilogram•meters per second east. (Neglect friction.)Calculate the magnitude of the cannon's velocity after the ball is fired. (Show all work, including the equation and subsitution with units.)

Part C

Base your answers to questions 1 and 2 on the information below:

A 1200-kilogram car moving at 12 meters per second collides with a 2300-kilogram car that is waiting at rest at a traffic light. After the collision, the cars lock together and slide. Eventually, the combined cars are brought to rest by a force of kinetic friction as the rubber tires slide across the dry, level, asphalt road surface.

1. Calculate the speed of the locked-together cars immediately after the collision. (Show all work, including the equation and substitution with units.)

2. Calculate the magnitude of the frictional force that brings the locked-together cars to rest. (Show all work, including the equation and substitution with units.)

Base your answers to questions 3 through 5 on the information below.
   A kicked soccer ball has an initial velocity of 25 meters per second at an angle of 40.° above the horizontal, level ground. (Neglect friction.)

3. Calculate the magnitude of the vertical component of the ball's initial velocity. (Show all work, including the equation and substitution with units.)

4. Calculate the maximum height the ball reaches above its initial position. (Show all work, including the equation and substitution with units.)

5. On the diagram, sketch the path of the ball's flight from its initial position at point *P* until it returns to level ground.

P   Level ground

Base your answers to questions 6 and 7 on the information and diagram below:

A pop-up toy has a mass of 0.020 kilogram and a spring constant of 150 newtons per meter. A force is applied to the toy to compress the spring 0.050 meter.

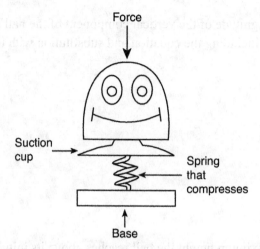

6. Calculate the potential energy stored in the compressed spring. (Show all work, including the equation and substitution with units.)

7. The toy is activated and all the compressed spring's potential energy is converted to gravitational potential energy. Calculate the maximum vertical height to which the toy is propelled. (Show all work, including the equation and substitution with units.)

Base your answers to questions 8 through 10 on the information below:

A car on a straight road starts from rest and accelerates at 1.0 meter per second^2 for 10. seconds. Then the car continues to travel at constant speed for an additional 20. seconds.

8. Determine the speed of the car at the end of the first 10. seconds.

_____m/s

9. On the grid below, use a ruler or straightedge to construct a graph of the car's speed as a function of time for the entire 30.-second interval.

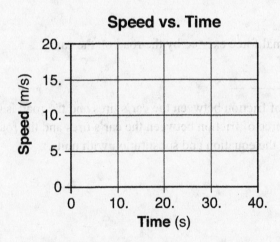

Speed vs. Time

10. Calculate the distance the car travels in the first 10. seconds. (Show all work, including the equation and substitution with units.)

Base your answers to questions 11 through 15 on the information below:
A manufacturer's advertisement claims that their 1,250-kilogram (12,300-newton) sports car can accelerate on a level road from 0 to 60.0 miles per hour (0 to 26.8 meters per second) in 3.75 seconds.

11. Determine the acceleration, in meters per second^2, of the car according to the advertisement.

_____ m/s^2

12. Calculate the net force required to give the car the acceleration claimed in the advertisement. (Show all work, including the equation and substitution with units.)

13. What is the normal force exerted by the road on the car?

_____ N

14. The coefficient of friction between the car's tires and the road is 0.80. Calculate the maximum force of friction between the car's tires and the road. (Show all work, including the equation and substitution with units.)

15. Using the values for the forces you have calculated, explain whether or not the manufacturer's claim for the car's acceleration is possible.

_____

_____

_____

PART II: WAVES, SOUND, AND LIGHT

UNIT 7
Waves, Sound, and Light Properties

Waves transfer energy through a *medium* or through empty space. There are many different examples of energy transfer through waves. This unit will explore the general properties of waves and overview the properties of sound and light waves.

Pulses and Periodic Waves

Imagine that you and a friend have a long rope that you hold between you. Your friend stays still and you give the rope a single quick snap up and back down with your hand. With this motion you will send a **pulse** down the length of the rope. The rope is the medium through which the pulse is traveling. A pulse like this is defined as a single vibratory disturbance in a medium.

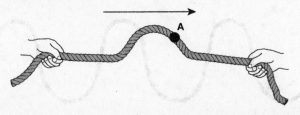

In the image above, you can see a pulse in a rope that is traveling to the right. What would happen to point A on the rope? This pulse is transferring energy along the rope; it is not moving matter along the rope. Therefore, point A will move up, then back down. Point A will NOT move to the right, however.

Now, imagine that you did not just send one pulse along the rope to your friend, but you kept repeating your motion. A series of pulses is defined as a **periodic wave**. If the end of the rope were to be moved up and down repeatedly, a wave pattern similar to a mathematical sine curve would appear in the rope. Each point in the rope would move in the vertical direction (up and down) and the wave energy would be *propagating* (transmitting) to the right, as in the diagram below.

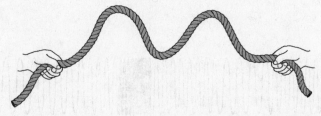

Each of the examples above represents wave energy traveling through a material medium. All waves that require a material medium through which to travel are defined as **mechanical waves**. Examples of mechanical waves include sound waves, water waves, and waves in ropes and springs. Waves that do not require a medium through which to travel are called **electromagnetic waves.** Some examples of electromagnetic waves are visible light, x-rays and radio waves. We will explore others later in this unit.

Types of Waves

There are two major types of waves: **transverse** and **longitudinal**. The type of wave is determined by the comparison of the direction of travel of the pulses to the direction of travel of individual particles of the medium.

Transverse waves are those in which the particles of the medium vibrate *perpendicular to* the direction of travel of the pulses. Transverse waves look like sine curves plotted on a graph. Light is the most obvious example of a transverse wave. Other examples include any electromagnetic wave, most waves in ropes and springs, and some seismic waves (S-waves). A stadium full of people doing "the wave" at a sporting event is a common human demonstration of a transverse wave.

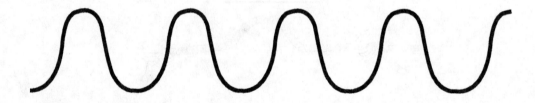

Longitudinal waves are those in which the particles of the medium vibrate *parallel to* the direction of travel of the pulses. Imagine taking a stretched slinky and moving the end of the slinky forward and back, rather than up and down like a transverse wave.

Sound is the most obvious example of a longitudinal wave. Other examples include compressional waves in springs, and some seismic waves (P-waves). A common human demonstration of a longitudinal wave involves lining up shoulder-to-shoulder and rocking to one side, bumping the person next to you, and then rocking back to your original position, with the process continuing down the line.

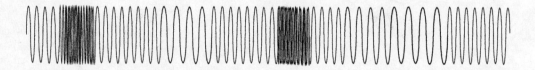

Parts of Waves

Whether the wave is transverse or longitudinal, it is important to know the different parts of the wave. Each part is described below.

- The **crest** of a transverse wave is the highest point from the rest position.

- The **trough** of a transverse wave is the lowest point from the rest position.

- A **compression, or condensation** in a longitudinal wave is an area of high pressure, or high density.

- A **rarefaction, or expansion** in a longitudinal wave is an area of low pressure, or low density.

- **Amplitude** is the maximum displacement of a particle in the medium from its rest (original) position, which corresponds to the energy of the wave. For a visible light wave, the amplitude is observed as brightness, or intensity of the light. This is the vertical distance from the rest position to a crest, or from the rest position to a trough. For a sound wave, the amplitude is observed as the loudness of the sound. On diagrams, the amplitude is usually labeled as A.

- The **wavelength** is the distance between two corresponding points on successive waves. The wavelength is most easily noted on transverse waves as the distance from one crest to the next, or one trough to the next. On longitudinal waves, it is the distance from one compression to the next, or one rarefaction to the next. On diagrams, the wavelength is labeled with the Greek letter *lambda* (λ).

The parts of transverse waves are labeled below:

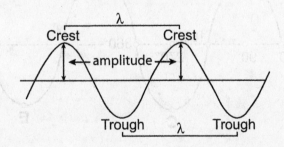

The parts of longitudinal waves are labeled below:

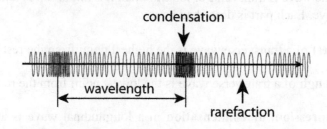

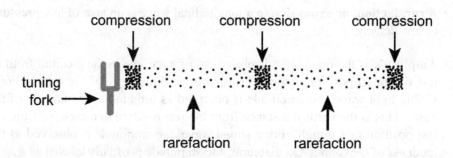

- **Phase:** Since a wave repeats itself each cycle, it can be represented like the degree measurements in a circle: a total of 360°. Corresponding points on successive waves are said to be *in phase* with each other when they are 360° apart. If the points are opposite each other (such as a crest and a trough), these points are said to be *out of phase,* specifically 180° out of phase with each other.

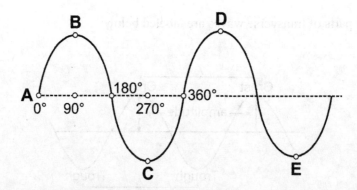

In diagram above, the following can be observed:
- points B and D are in phase
- points C and E are in phase
- points B and C are 180° out of phase
- points D and E are 180° out of phase

Periodic Wave Phenomena

The following are properties of all periodic waves.

- The **frequency (f)** of a wave represents the number of complete waves, or cycles, passing a given point per second. The unit of frequency is cycles per second, or s^{-1}. This is also typically represented as the hertz (Hz), named for *Heinrich Hertz*. For a visible light wave, the frequency is observed as a particular color of the light. For a sound wave, the frequency is observed as the pitch of the sound.

- The **period (T)** of a wave is the time that it takes a wave to complete one cycle. Since period is a measurement of time, the unit of period is the second(s). Frequency and period are related to each other in that they are reciprocals, as seen in the equation below:

$$f = \frac{1}{T} \text{ or } T = \frac{1}{f}$$

- **Speed** (v) of a wave is dependent on the type and characteristics of the medium through which it is traveling. If characteristics of the medium, such as temperature and tension, do not change, then the wave will travel at a constant speed. Wave speed can be calculated in a similar fashion to mechanical speed, using the following equation:

$$v = \frac{d}{t}$$

Speed can also be represented in terms of wavelength (λ) and frequency (f):

$$v = f\lambda$$

Wave Interactions

When a pulse or a wave interacts with a boundary, one of a variety of things may occur. The pulse could **reflect, transmit, or be absorbed**. Reflection is the bouncing of a pulse or wave off of the boundary. Transmission is the passing of a pulse or wave through the boundary. Absorption is the conversion of the wave energy into some other form.

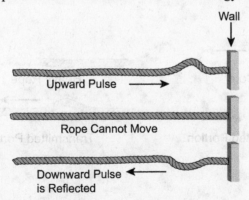

In the diagram above, an upward pulse is sent along a rope toward a wall where the end of the rope is fixed in position. When the pulse hits this fixed end at the wall, the pulse is **reflected** off of the wall and it is *inverted* (upside down) as compared to the original pulse. In this example the inversion of the pulse causes a *change of phase* of the pulse from a crest to a trough. When pulses reflect off of a fixed end, the pulse will be inverted.

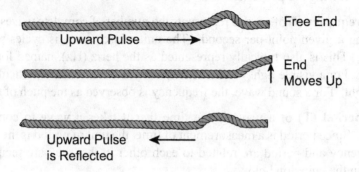

In the diagram above, an upward pulse is sent along a rope toward an end that is not fixed and is free to move. In this situation, the pulse is **reflected** off of this free end and it keeps the same phase orientation (it is NOT inverted).

When a pulse reaches a boundary of one material to another, a combination of the interactions will occur. Some of the energy is transmitted into the other rope or cord, and some of the energy is reflected back in the opposite direction, as seen in both diagrams below.

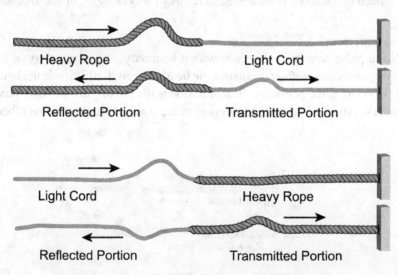

The Doppler Effect

If the wave source and receiver are in motion relative to each other, the frequency will be different than the original source frequency. This phenomenon is called the **Doppler Effect**, named for Christian Doppler. The Doppler Effect occurs for any wave but is most noticeable with sound and light.

Christian Doppler

Christian Doppler was an Austrian physicist, mathematician, and philosopher most known for his explanation of the Doppler Effect. In his description, he postulated that the observed frequency of a wave depends on the relative speed of the source and the observer, which explained the color shift of stars.

Born: November 29, 1803; Austria
Died: March 17, 1853; Italy

Colleges Attended:
Vienna Polytechnic Institute
University of Vienna

If the source of a wave is moving closer to the receiver, the receiver will observe a frequency of that wave that is higher than the source frequency. If the source is moving away from the receiver, the receiver will observe a frequency that is lower than the original source frequency.

The Doppler Effect actually occurs because the *wavelength* of the wave is either compressed (source moving toward receiver), or stretched out (source moving away). Since wavelength and frequency are inversely related, the frequency is shifted as well. For sound, the observed effect is a higher pitch sound if the source is moving closer to you, and a lower pitch sound if the source is moving away. The Doppler Effect for sound is especially noticeable in emergency vehicle sirens, car horns, and racing car engines. An example of this is shown in the diagram on the following page.

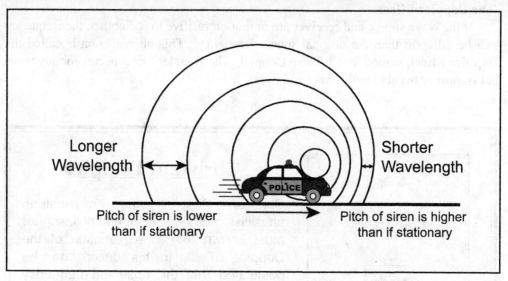

For light, there is a shift in the color of the light. If the source is moving toward you, a normally white light would be shifted to appear to be blue in color. This is often referred to as *blue shift*. Likewise, if the source is moving away from you, the normally white source would appear red in color and is referred to as *red shift*. The Doppler Effect for light is observable in space with the red shift of the stars. The fact that stars may appear red in color is evidence that the universe is expanding and these stars are moving farther away from Earth. This is shown in the diagram below.

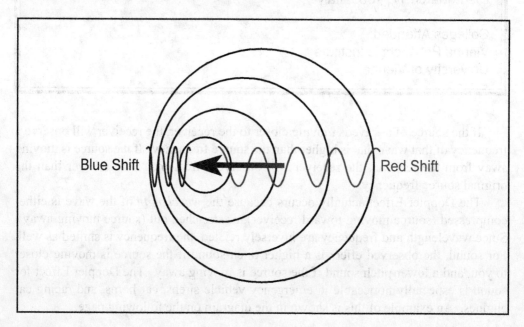

Superposition

When two pulses or waves are in the same place at the same time, they will combine their energies in a specific way. This is called **superposition**.

When two pulses or waves are *in phase* with respect to each other (for example, crests lined up with crests) maximum **constructive interference** occurs. This will result in a wave that has the combined amplitudes of the individual waves, as shown in the diagram below:

When two pulses or waves are *out of phase* with respect to each other (the crests are lined up with troughs) maximum **destructive interference** occurs. If the amplitudes of the individual waves are equal, total destructive interference occurs and the result is an amplitude of zero. Destructive interference will result in a wave that has the difference between the amplitudes of the individual waves, as shown in the diagram below:

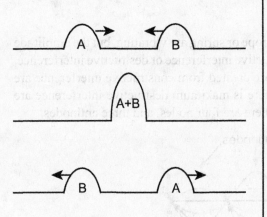

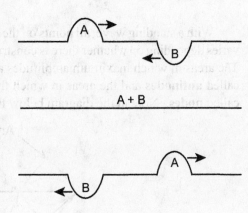

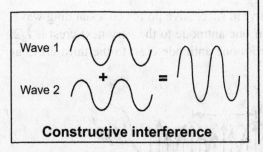

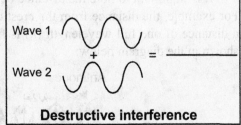

Constructive interference

Destructive interference

Standing Waves

When two waves of the same frequency and amplitude continuously pass through each other traveling in opposite directions, **standing waves** will be produced. The standing wave is the principle behind the vibrations of stringed instruments, such as a guitar. They can easily be demonstrated using a long rope or spring stretched between two people. If the person on one end continuously shakes the spring or rope up and down at regular intervals, the pulses will travel to the other end, reflect back inverted, and combine with the incoming pulses. The resulting pattern will resemble the diagram below:

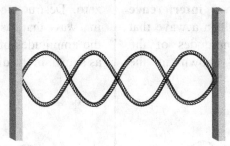

With a standing wave, all points on the rope or spring are vibrating, but the amplitude varies depending on whether there is constructive interference or destructive interference. The areas in which maximum amplitudes are created from constructive interference are called **antinodes** and the areas in which there is maximum destructive interference are called **nodes**. Note in the diagram below there are four nodes, and three antinodes.

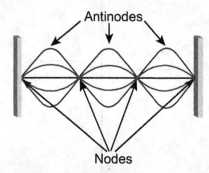

It is important to note the distance between successive points on a standing wave. For example, the distance from the crest of one antinode to the very next crest is $\lambda/2$; a distance of one full wavelength (λ) is from one antinode crest to the third crest, as shown in the diagram below:

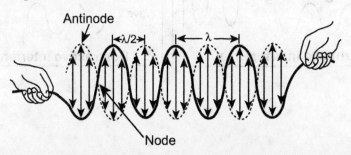

WAVES, SOUND, AND LIGHT PROPERTIES

If the experiment were conducted using a long spring and a person were to shake this spring at a different frequency, there would be a different number of nodes and antinodes. For a given length of rope or spring, L, different *harmonics* are labeled below with the corresponding diagrams of nodes and antinodes.

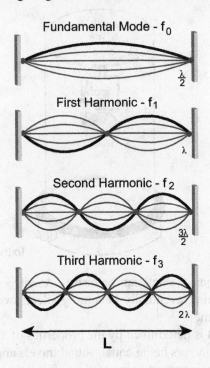

Fundamental Mode - f_0

$\frac{\lambda}{2}$

First Harmonic - f_1

λ

Second Harmonic - f_2

$\frac{3\lambda}{2}$

Third Harmonic - f_3

2λ

L

Properties of Sound

Sound is a *longitudinal mechanical wave,* which is the result of a vibrating medium. In order for a person to hear a sound, the air around a person's ear must vibrate which causes the eardrum and the inner ear to vibrate, which is then detected by the nerves and interpreted by the brain.

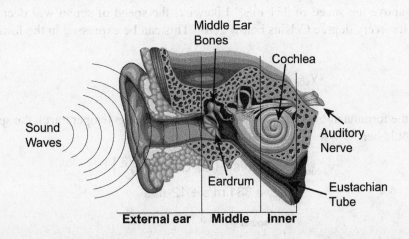

Middle Ear Bones

Cochlea

Sound Waves

Auditory Nerve

Eardrum

Eustachian Tube

External ear   **Middle**   **Inner**

Sound requires a material medium through which to travel. Sound is most often associated with traveling through the air, but it can also travel through solids, liquids, and other gases. Sound can not travel through a vacuum. This can easily be demonstrated by using a vacuum jar with a mechanical bell inside.

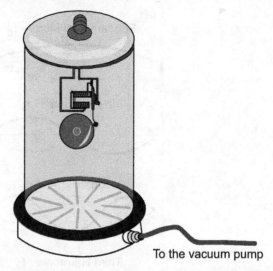

To the vacuum pump

With air inside the jar, the bell can be heard. As air is evacuated, the sound will grow more and more faint, eventually becoming silent, even though the bell is still being struck by the hammer.

The **speed of sound** is determined by the properties of the medium through which it is traveling. All other factors being equal, sound travels approximately 4 times faster through water than through air, and about 15 times faster through steel than through air. At STP (standard temperature and pressure), the speed of sound in air is 331 m/s, as shown in the Reference Tables. As the characteristics of the air change (such as pressure, temperature, humidity, and impurities), the speed of sound will vary. If nothing else is to change except temperature of the air, the speed of sound is directly proportional to the temperature. For every degree Celsius above zero, the speed of sound will increase by 0.6 m/s above the speed of 331 m/s. Likewise, the speed of sound will decrease by 0.6 m/s for every degree Celsius below zero. This can be expressed in the formula below:

$$V_{sound} = 331 m/s + T\left(0.6 \ \frac{m/s}{°C}\right)$$

Using the formula above, for a temperature of 20°C (room temperature), the speed of sound would be:

$$V_{sound} = 331 \ m/s + (20°C)\left(0.6 \frac{m/s}{°C}\right)$$

$$V_{sound} = 331 \ m/s + 12 \ m/s$$

$$V_{sound} = 343 \ m/s$$

WAVES, SOUND, AND LIGHT PROPERTIES

The amplitude of a sound wave is the loudness, or volume level of a sound. Loudness of a sound is measured with a logarithmic scale called the decibel scale. Some everyday examples of sound levels on the decibel scale are below:

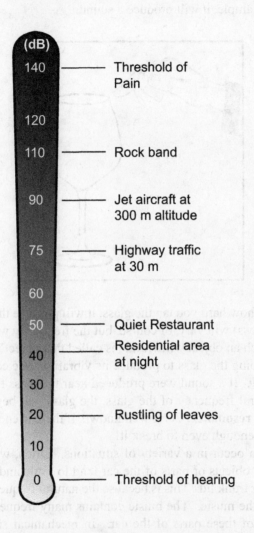

(dB)	
140	Threshold of Pain
120	
110	Rock band
90	Jet aircraft at 300 m altitude
75	Highway traffic at 30 m
60	
50	Quiet Restaurant
40	Residential area at night
30	
20	Rustling of leaves
10	
0	Threshold of hearing

The frequency of sound is the pitch that you hear. For a person with good hearing, the typical range of discernible frequencies is 20 Hz - 20,000 Hz. Most sounds are not a single frequency, but are rather a combination of many frequencies. As a person ages, and depending on exposure to loud sounds during one's lifetime, the ability to hear the entire range diminishes, especially in the upper frequencies. This is why older people can't easily hear very high-pitched sounds. Other animals, such as dogs, have a higher upper limit for frequency detection. This is why a dog whistle is effective; humans can't hear it, but dogs can.

Resonance

Since a sound is produced by a vibrating object, one can determine the sound an object will produce simply by making it vibrate. If you were to tap a wine glass with your fingernail, for example, it will produce a sound.

Regardless of how hard you tap the glass, it will produce the same pitch sound. The amplitude (loudness) will vary, of course, but the frequency will remain the same. The frequency at which an object will vibrate is called that object's *natural frequency*.

Instead of tapping the glass to produce its vibration, one could get the glass to vibrate in another way. If a sound were produced near the glass that was of the same frequency as the natural frequency of the glass, the glass will begin to vibrate. This phenomenon is called **resonance**. As is well known, if there is enough energy present, the glass may vibrate enough even to break it!

Resonance can occur in a variety of situations. Often, when listening to the radio in the car, other objects or parts of the car tend to rattle and vibrate such as the mirror, door panels, or trunk lid. This is because the natural frequency of these objects is being produced in the music. The music contains many frequencies, including the resonant frequencies of these parts of the car. In mechanical situations, resonance occurs when pushing someone on a playground swing, as you must push the swing at its resonant frequency in order to swing higher (increase amplitude).

Properties of Light

Light is a transverse electromagnetic wave and does not require a medium through which to travel. The most obvious example of light is *visible light* but there are many other electromagnetic waves that can also be labeled as light. These other electromagnetic waves are not visible to the naked eye. All of the transverse electro-magnetic waves - visible or not - make up **the electromagnetic (EM) spectrum.**

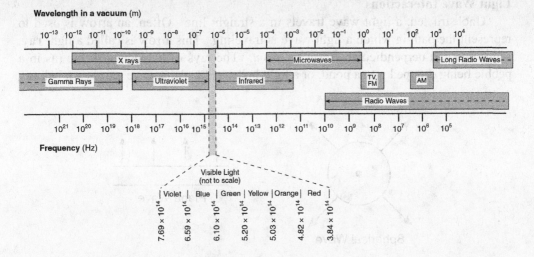

The chart above, from the Reference Tables, lists the major components of the electromagnetic spectrum. The different components of the EM spectrum are identified by their frequency. From left to right on the chart, the frequency decreases (and the wavelength increases) as noted on the scales. There are some areas where a certain frequency is represented by two types of EM waves. At these frequencies, the wave exhibits properties of each type of wave. The energy of an electromagnetic wave is directly proportional to the frequency, so the most energetic (and likewise, most damaging to one's health) are found on the left side of the chart.

The amplitude of a light wave determines that wave's intensity. For visible light, this intensity is observed as brightness.

Speed of Light

All electromagnetic waves travel at the same speed *through a vacuum*. That speed has been measured and refined many times in history. Galileo described an experiment that involved uncovering lanterns at a distance of about one mile, but this technique was never successful. An astronomer named Olaf Roemer in 1676 used the eclipses of one of the moons of Jupiter and Earth's revolution to determine the speed of light to be approximately 186,000 miles per second. The current accepted value for the speed of light is 186,282 miles per second, or 299,792,458 meters per second. This is referred to as c and, rounded to 3 significant figures, is defined as:

$$c = 3.00 \times 10^8 \text{ m/s}$$

The speed of light in a material medium is always less than its speed in a vacuum. It varies with *optical density* of the medium. The speed of light in air is very close to the speed of light in a vacuum. For most practical purposes, it may be considered to be the same.

The relationship between light's speed, frequency, and wavelength is the same as any other wave:

$$v = f\lambda$$

133

Light Wave Interactions

Unobstructed, a light wave travels in a straight line. Often, an arrow is used to represent the path in which a light wave is traveling. This arrow is called a light **ray**. Light rays are perpendicular to a *wave front*. The rays of a spherical wave (as in a pebble being dropped into a pond, or a light bulb) and a plane wave are shown below:

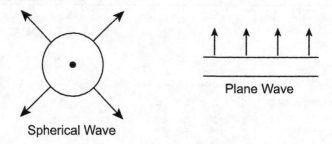

Spherical Wave

Plane Wave

Light falling on an object may be *absorbed, transmitted, or reflected*, depending on the characteristics of the object. For example, if light were to fall onto the roof of a car, some of that light would reflect off of the surface and some of the light energy would be absorbed, but none will pass through. This is because the roof of a car is an *opaque* object - one that does not let light transmit. If the car were white in color, much of the light would be reflected as compared to a black car that would absorb much of the light energy. A sheet of paper would transmit some of the light but the light rays would be interrupted and scattered as they pass. This is because the paper is *translucent*. Light rays do not pass straight through translucent materials and you can't see right through them. Some of the light is absorbed, and some is reflected off of the paper. A *transparent* material, like a clear glass window, will allow most of the light to pass right through uninterrupted.

Objects that give off their own light are said to be *luminous* objects. Examples of luminous objects include lit light bulbs, the sun, and lit LEDs. Objects that are visible due to the fact that light is reflecting off of them are *illuminated* objects. Examples of illuminated objects include the moon, a table, and this book.

Polarization

The polarization of a wave refers to the orientation of the vibration. To use a mechanical example, imagine a long rope stretched between two people. If one of the people were to produce a transverse wave by shaking the rope up and down, that wave is described as being vertically polarized. On the other hand, the person could also produce a transverse wave by shaking the rope from side to side. Then, the waves would be horizontally polarized. Or, the person could shake the rope at any other angle to produce the wave.

Normally, light waves are **unpolarized**. That means that continuous light waves have all different orientations, rather than just one. The way to polarize these light waves is to pass them through a *polarizing filter*. Mechanically speaking, this is similar to passing the rope through an opening, as shown in the diagram below:

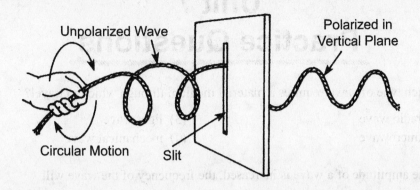

Passing light (or, mechanically speaking, rope) through two polarizing filters yields the results below:

The Picket Fence Analogy

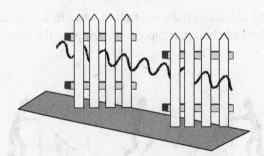

When the pickets of both fences are aligned in
the vertical direction, a vertical vibration can
make it through both fences.

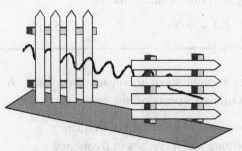

When the pickets of the second fence are horizontal,
vertical vibrations which make it through the first
fence will be blocked.

Unit 7
Practice Questions

1. Which type of wave requires a material medium through which to travel?

   (1) radio wave
   (2) microwave
   (3) light wave
   (4) mechanical wave

2. If the amplitude of a wave is increased, the frequency of the wave will

   (1) decrease   (2) increase   (3) remain the same

3. Which unit is equivalent to meters per second? $Hz = S^{-1}$

   (1) Hz•s   (2) Hz•m   (3) s/Hz   (4) m/Hz

4. While playing, two children create a standing wave in a rope, as shown in the diagram below. A third child participates by jumping the rope.

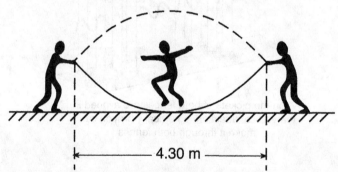

   — 4.30 m —

   What is the wavelength of this standing wave?

   (1) 2.15 m   (2) 4.30 m   (3) 6.45 m   (4) 8.60 m

5. A car's horn produces a sound wave of constant frequency. As the car speeds up going away from a stationary spectator, the sound wave detected by the spectator

   (1) decreases in amplitude and decreases in frequency
   (2) decreases in amplitude and increases in frequency
   (3) increases in amplitude and decreases in frequency
   (4) increases in amplitude and increases in frequency

6. Approximately how much time does it take light to travel from the Sun to Earth?

(1) 2.00×10^{-3} s

(2) 1.28×10^0 s

(3) 5.00×10^2 s

(4) 4.50×10^{19} s

7. The time required for a wave to complete one full cycle is called the wave's

(1) frequency

(2) period

(3) velocity

(4) wavelength

8. The diagram below represents a transverse wave

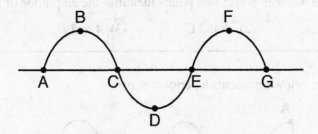

The wavelength of the wave is equal to the distance between points

(1) A and G

(2) B and F

(3) C and E

(4) D and F

9. The diagram below represents a transverse wave traveling to the right through a medium. Point A represents a particle of the medium.

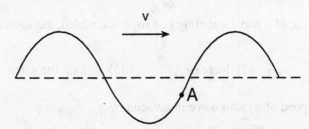

In which direction will particle A move in the next instant of time?

(1) up

(2) down

(3) left

(4) right

10. What is the period of a 60.-hertz electromagnetic wave traveling at 3.0×10^8 meters per second?

(1) 1.7×10^{-2} s

(2) 2.0×10^{-7} s

(3) 6.0×10^1 s

(4) 5.0×10^6 s

11. The diagram below represents a transverse wave.

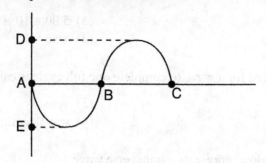

The distance between which two points identifies the amplitude of the wave?

(1) *A* and *B*   (2) *A* and *C*   (3) *A* and *E*   (4) *D* and *E*

---

12. The diagram below represents a periodic wave.

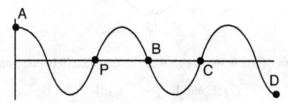

Which point on the wave is in phase with point P?

(1) *A*   (2) *B*   (3) *C*   (4) *D*

---

13. If the amplitude of a wave traveling in a rope is doubled, the speed of the wave in the rope will

(1) decrease   (2) increase   (3) remain the same

14. What is the speed of a radio wave in a vacuum?

(1) 0 m/s   (3) 1.13×10^3 m/s
(2) 3.31×10^2 m/s   (4) 3.00×10^8 m/s

15. A ringing bell is located in a chamber. When the air is removed from the chamber, why can the bell be seen vibrating but not be heard?

(1) Light waves can travel though a vacuum, but sound waves cannot.
(2) Sound waves have greater amplitude than light waves.
(3) Light waves travel slower than sound waves.
(4) Sound waves have higher frequency than light waves.

16. Which wavelength is in the infrared range of the electromagnetic spectrum?

    (1) 100 nm   (2) 100 mm   (3) 100 m   (4) 100 µm

17. The diagram below represents a wave.

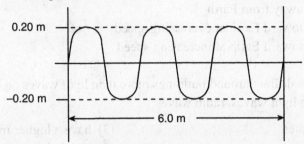

What is the speed of the wave if its frequency is 8.0 hertz?

    (1) 48 m/s   (2) 16 m/s   (3) 3.2 m/s   (4) 1.6 m/s

18. A person observes a fireworks display from a safe distance of 0.750 kilometer. Assuming that sound travels at 340. meters per second in air, what is the time between the person seeing and hearing a fireworks explosion?

    (1) 0.453 s   (3) 410. s
    (2) 2.21 s   (4) 2.55×10^5 s

19. Electromagnetic radiation having a wavelength of 1.3×10^{-7} meter would be classified as

    (1) infrared   (2) orange   (3) blue   (4) ultraviolet

20. The diagram below represents a transverse wave traveling in a string.

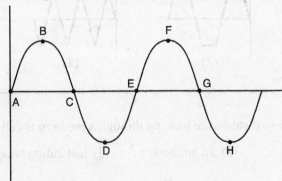

Which two labeled points are 180° out of phase?

    (1) A and D   (2) B and F   (3) D and F   (4) D and H

21. When observed from Earth, the wavelengths of light emitted by a star are shifted toward the red end of the electromagnetic spectrum. This redshift occurs because the star is

    (1) at rest relative to Earth
    (2) moving away from Earth
    (3) moving toward Earth at decreasing speed
    (4) moving toward Earth at increasing speed

22. Radio waves diffract around buildings more than light waves do because, compared to light waves, radio waves

    (1) move faster   (3) have a higher frequency
    (2) move slower   (4) have a longer wavelength

23. A tuning fork vibrating in air produces sound waves. These waves are best classified as

    (1) transverse, because the air molecules are vibrating parallel to the direction of wave motion
    (2) transverse, because the air molecules are vibrating perpendicular to the direction of wave motion
    (3) longitudinal, because the air molecules are vibrating parallel to the direction of wave motion
    (4) longitudinal, because the air molecules are vibrating perpendicular to the direction of wave motion

24. Which diagram below does not represent a periodic wave?

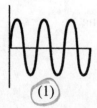 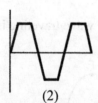 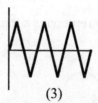 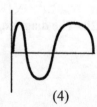

    (1)   (2)   (3)   (4)

25. A single vibratory disturbance moving through a medium is called

    (1) a node   (2) an antinode   (3) a standing wave   (4) a pulse

26. If the frequency of a periodic wave is doubled, the period of the wave will be

    (1) halved   (2) doubled   (3) quartered   (4) quadrupled

27. Which wave diagram has both wavelength (λ) and amplitude (A) labeled correctly?

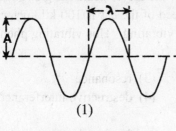

(1)

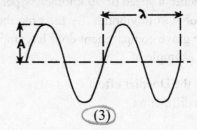

(3)

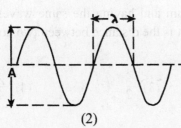

(2)

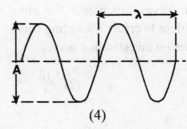

(4)

28. A periodic wave transfers

    (1) energy, only   (3) both energy and mass

    (2) mass, only   (4) neither energy nor mass

29. A motor is used to produce 4.0 waves each second in a string. What is the frequency of the waves?

    (1) 0.25 Hz   (2) 15 Hz   (3) 25 Hz   (4) 4.0 Hz

30. The diagram below shows two points, A and B, on a wave train

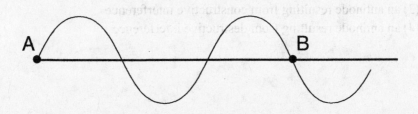

How many wavelengths separate point A and point B?

    (1) 1.0   (2) 1.5   (3) 3.0   (4) 0.75

31. A car traveling at 70 kilometers per hour accelerates to pass a truck. When the car reaches a speed of 90 kilometers per hour the driver hears the glove compartment door start to vibrate. By the time the speed of the car is 100 kilometers per hour, the glove compartment door has stopped vibrating. This vibrating phenomenon is an example of

(1) the Doppler effect

(2) diffraction

(3) resonance

(4) destructive interferance

32. Two waves traveling in the same medium and having the same wavelength (λ) interfere to create a standing wave. What is the distance between two consecutive nodes on this standing wave?

(1) λ

(2) $\dfrac{3\lambda}{4}$

(3) $\dfrac{\lambda}{2}$

(4) $\dfrac{\lambda}{4}$

33. The diagram below shows a standing wave.

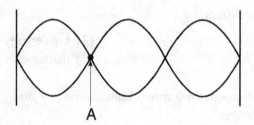

A

Point A on the standing wave is

(1) a node resulting from constructive interference

(2) a node resulting from destructive interference

(3) an antinode resulting from constructive interference

(4) an antinode resulting from destructive interference

UNIT 8
Reflection and Refraction

Reflection

When light rays fall on a surface and bounce off of it, **reflection** occurs. All waves will reflect, but it is most apparent in visible light. The rays will follow the **Law of Reflection** which states that the *angle of incidence* of a light ray measured with respect to a normal (perpendicular) line is equal to the *angle of reflection*. This is shown in the equation and diagram below:

$$\theta_i = \theta_r$$

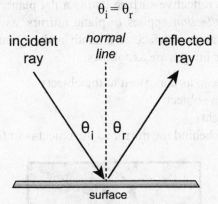

All points on a reflecting surface follow the law of reflection. However, it may not be noticeable on a large scale due to the relative roughness of the surface. When the surface is smooth and flat, such as a plane mirror, parallel incident rays of light will reflect off of the surface and remain parallel to each other. This is called **regular** (or **specular**) **reflection**. When a reflecting surface is irregular or rough, such as a piece of shiny steel, parallel incident rays of light will reflect off of the surface and be scattered, which is called diffuse reflection. However, for each individual ray of light, the law of reflection still applies.

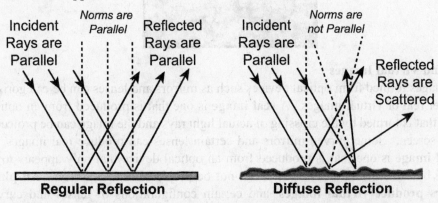

Plane Mirrors

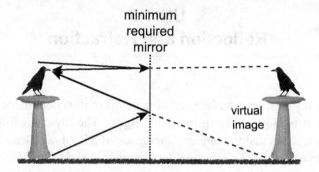

A **plane mirror** is a reflective surface that is a flat plane, as compared to curved mirrors. The *Law of Reflection* applies to plane mirrors, as it does for all types of reflection. Assuming that the surface is smooth and regular reflection occurs, the properties of plane mirror images are as follows:

- upright (right side up as compared to the object)
- the same size as the object
- reversed left to right
- the same distance behind the mirror as the object is in front
- virtual

Real and Virtual Images

Images formed from optical devices such as mirrors and lenses can be categorized as either real or virtual images. A **real image** is one that is produced from an optical device that is formed by the crossing of actual light rays and the image can be projected onto a screen. Some curved mirrors and certain lenses can produce real images. A **virtual image** is one that is produced from an optical device that only appears to be formed by crossing light rays and it can not be projected onto a screen. **All plane mirrors produce virtual images**, and certain configurations of lenses and curved mirrors can produce virtual images as well.

Refraction

Light rays that are incident on another medium can reflect, absorb, or transmit, depending on the properties of the material. If the material is optically transparent, some of the light energy may reflect and absorb, but much of it will transmit, or pass into the other material. When this light enters a new medium, the speed of that light will change. Light travels the fastest through a vacuum, and only slightly slower through air. Through any other media, light travels more slowly. The property of a material that describes the speed of light through it is the **index of refraction (n)**. The index of refraction (n) for a material represents the ratio of the speed of light through a vacuum (c) to its speed through that material (v). The index of refraction is a representation of the *optical density* of a substance.

$$n = \frac{c}{v}$$

Since light travels fastest through a vacuum, the value of n will always be larger than one. Likewise, v will always be less than c, which is 3.00×10^8 m/s. From the Reference Tables, the values of the indices of refraction for a variety of materials are below:

Absolute Indices of Refraction $(f = 5.09 \times 10^{14}$ Hz$)$	
Air	1.00
Corn oil	1.47
Diamond	2.42
Ethyl alcohol	1.36
Glass, crown	1.52
Glass, flint	1.66
Glycerol	1.47
Lucite	1.50
Quartz, fused	1.46
Sodium chloride	1.54
Water	1.33
Zircon	1.92

In the chart above, it is important to note that the frequency of light used to experimentally determine these values of the indices of refraction is 5.09×10^{14} Hz. From the electromagnetic spectrum, this is a yellow colored light. If another frequency of light were used, the indices of refraction for these materials would be different.

Looking at the formula for the index of refraction, it can be noted that the index of refraction is inversely proportional to the speed of light through that material. Therefore, from the chart, the material listed through which light travels most slowly is diamond.

145

The wave equation $(v=f\lambda)$ applies to light entering a new medium. Since the speed of light changes as it enters this new medium and the frequency stays the same, the wavelength must then change as well. From the reference tables, the relationship among index of refraction, speed, and wavelength of light for two different media is below:

$$\frac{n_2}{n_1} = \frac{v_1}{v_2} = \frac{\lambda_1}{\lambda_2}$$

If a light ray enters another medium from air at an *oblique* angle (any angle other than perpendicular to the surface), not only will the light ray slow down, but it will also bend upon entry and change its direction. This change in direction is mathematically described by **Snell's Law**.

$$n_1 \sin\theta_1 = n_2 \sin\theta_2$$

Willebrord Snell

Willebrord Snell was a Dutch astronomer and mathematician best known for Snell's Law, which states that transparent materials have different indices of refraction. This relationship is attributed to Snell although it has since been linked to other scientists many years earlier. Snell also determined a new method of measuring the radius of the earth using triangulation.

Born: June 13, 1580; the Netherlands
Died: October 30, 1626; the Netherlands

One way to remember where to place certain data for a Snell's Law problem is to think that a subscript of 1 refers to the medium from which that the light is coming, and a subscript of 2 refers to the medium into which the light is going. An example of light traveling from air into water is below:

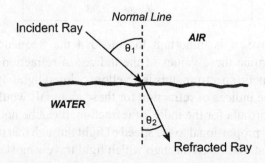

REFLECTION AND REFRACTION

In the previous example, the angle in the water is smaller than the angle in the air. When light passes from a substance of relatively low index of refraction into a substance of higher index of refraction, the light ray *bends towards the normal line*. Likewise, when a light ray passes from relatively high index of refraction to lower index of refraction, the light ray *bends away from the normal line*, as in the example shown to the right:

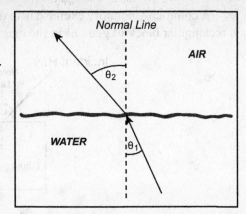

If light passes into another substance that has the same index of refraction as the original material, the light ray will not bend. According to the indices of refraction chart, corn oil and glycerol have the same index of refraction. If light were to go from the corn oil to the glycerol, it would not change speed nor would it bend, as in the diagram shown to the right:

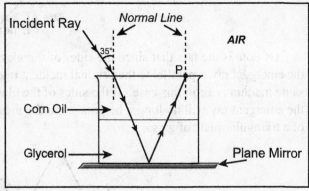

An interesting demonstration can be done to show this. Corn oil and the type of glass used to make beakers have essentially the same index of refraction. If a small beaker is placed inside a larger one and the beakers are filled with corn oil, the inner beaker seems to disappear as it is filled!

A common laboratory exercise is to diagram the refraction of light passing through a rectangular block of glass as in the diagram below:

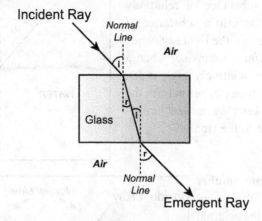

Incident Ray

Normal Line

Air

Glass

Air

Normal Line

Emergent Ray

Of note is the fact that since the sides of the glass block are parallel to each other, the emergent ray is parallel to the original incident ray, as long as the two rays are in the same medium - air in this case. If the sides of the glass plate are not parallel, however, the emergent ray will no longer be parallel to the incident ray, as in the diagram below of a triangular plate of glass:

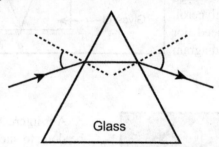

Glass

According to Snell's Law, the angle of refraction (θ_2) is directly proportional to the angle of incidence (θ_1). If light were to travel from a more optically dense material to a less optically dense material (from high index of refraction to lower index), the light ray bends away from the normal line. As the angle of incidence increases, so does the angle of refraction. At a certain point, the angle of refraction will become 90°. The specific angle of incidence for which the angle of refraction is 90° is called the **critical angle (θ_c)**.

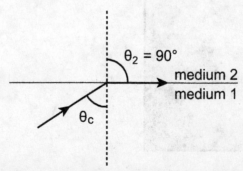

$\theta_2 = 90°$

medium 2
medium 1

θ_c

REFLECTION AND REFRACTION

The formula for the critical angle is an application of Snell's Law:

$$n_1 \sin \theta_1 = n_2 \sin \theta_2$$

$$n_1 \sin \theta_c = n_2 \sin(90°)$$

$$\sin \theta_c = \frac{n_2}{n_1}$$

In many cases, the second medium (n_2) is air, which has an index of refraction of 1. Therefore, the formula for the critical angle becomes:

$$\sin \theta_c = \frac{1}{n}$$

If the angle of incidence continues to increase beyond the critical angle, the light will no longer refract out of the medium, but it will now reflect back into the first medium. This is called **total internal reflection**. The boundary is acting like a mirror and the law of reflection applies.

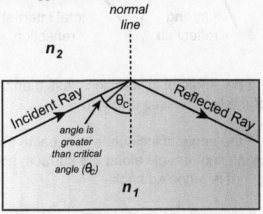

Total internal reflection is the principle behind fiber optics.

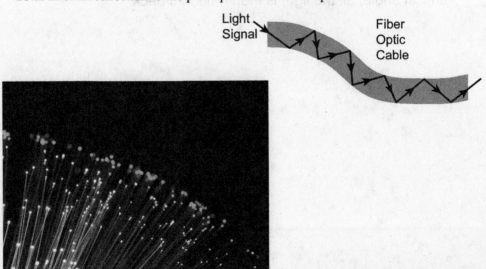

The following diagrams summarize the effect on increasing the angle of incidence:

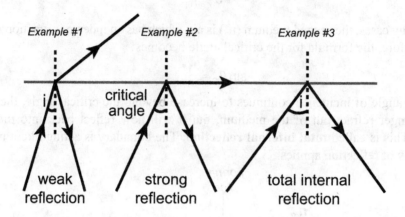

Example #1: If the angle of incidence is less than the critical angle, most of the light passes through.

Example #2: If the angle of incidence is equal to the critical angle, most of the light travels along the surface, however a significant amount is reflected back.

Example #3: If the angle of incidence is greater than the critical angle, all the light is then reflected back.

Unit 8
Practice Questions

1. What is the speed of light ($f = 5.09 \times 10^{14}$ Hz) in flint glass?

$n = \dfrac{c}{v}$

   (1) 1.81×10^8 m/s   (3) 3.00×10^8 m/s $1.66 = \dfrac{3 \times 10^8}{v}$

   (2) 1.97×10^8 m/s   (4) 4.98×10^8 m/s

$v =$

2. A ray of light ($f = 5.09 \times 10^{14}$ Hz) traveling in air is incident at an angle of $40.°$ on an air-crown glass interface as shown below.

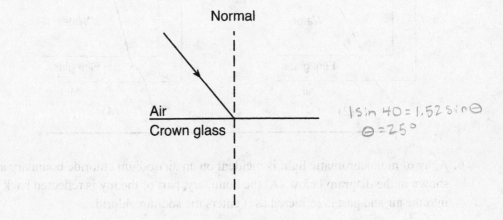

$1 \sin 40 = 1.52 \sin \theta$

$\theta = 25°$

What is the angle of refraction for this light ray?

   (1) $25°$ (2) $37°$ (3) $40°$ (4) $78°$

---

3. As yellow light ($f = 5.09 \times 10^{14}$ Hz) travels from zircon into diamond, the speed of the light

   (1) decreases   (2) increases   (3) remains the same

4. What is the wavelength of a light ray with frequency 5.09×10^{14} hertz as it travels through Lucite?

   (1) 3.93×10^{-7} m   (3) 3.39×10^{14} m $\dfrac{1}{1.5} = \dfrac{v}{3 \times 10^8}$

   (2) 5.89×10^{-7} m   (4) 7.64×10^{14} m

$v = 2 \times 10^8$

$\dfrac{2 \times 10^8}{3 \times 10^8} = 5.09 \times 10^{14} \lambda$

151

5. Which diagram best represents the path taken by a ray of monochromatic light as it passes from air through the materials shown?

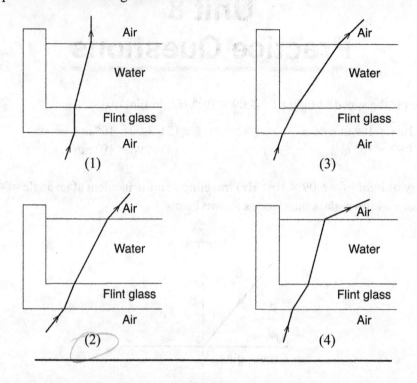

6. A ray of monochromatic light is incident on an air-sodium chloride boundary as shown in the diagram below. At the boundary, part of the ray is reflected back into the air and part is refracted as it enters the sodium chloride.

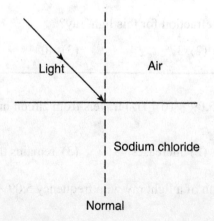

Compared to the ray's angle of refraction in the sodium chloride, the ray's angle of reflection in the air is

(1) smaller   (2) larger   (3) the same

7. A laser beam is directed at the surface of a smooth, calm pond as represented in the diagram below.

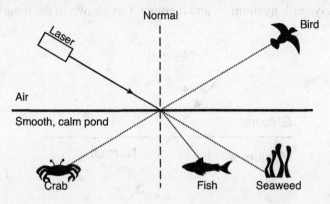

Which organisms could be illuminated by the laser light?

(1) the bird and the fish   (3) the crab and the seaweed
(2) the bird and the seaweed   (4) the crab and the fish

Base your answers to questions 8 and 9 on the diagram below, which represents a light ray traveling from air to Lucite to medium Y and back into air.

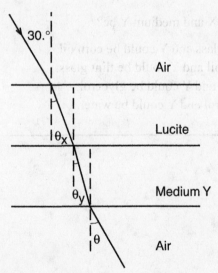

8. The sine of angle θ_x is

(1) 0.333   (2) 0.500   (3) 0.707   (4) 0.886

9. Light travels slowest in

(1) air, only   (3) medium Y, only
(2) Lucite, only   (4) air, Lucite, and medium Y

153

10. A beam of monochromatic light (f = 5.09 × 10¹⁴ hertz) passes through parallel sections of glycerol, medium X, and medium Y as shown in the diagram below.

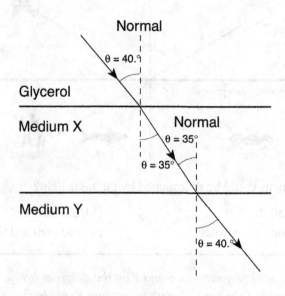

What could medium X and medium Y be?

(1) X could be flint glass and Y could be corn oil.
(2) X could be corn oil and Y could be flint glass.
(3) X could be water and Y could be glycerol.
(4) X could be glycerol and Y could be water.

UNIT 9
Diffraction and Interference

Recall from Unit 7 that two waves that are at the same place at the same time can *interfere* with each other. How they combine will determine whether it produces a wave with larger amplitude from *constructive interference,* or a wave with smaller amplitude from *destructive interference*.

Diffraction

In addition to some of the other properties of waves, it was discovered that waves will spread out when passing through an opening or past a barrier. This spreading of a wave is called **diffraction**. Diffraction can be easily demonstrated with parallel water waves passing through a small opening.

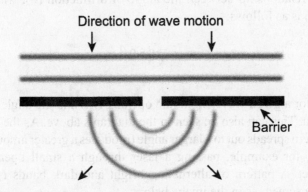

Direction of wave motion

Barrier

As shown in the diagram above, the wave fronts will spread out as they pass through the opening in the barrier. This diffraction is especially apparent when the size of the opening is of a similar order of magnitude as compared to the wavelength of the wave passing through it. With a very wide opening as compared to the wavelength, the pattern is less obvious.

The diagram below shows the differences in the pattern with small and large openings:

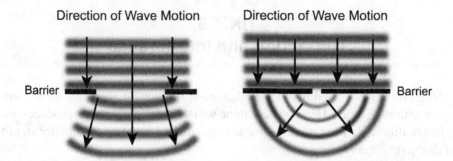

Diffraction occurs for all types of waves. Water waves, as seen in the diagrams above, give a visual representation of the pattern. Diffraction can also be observed with sound waves. For example, one can hear a person talking around the corner of a building, or off to the side through an open door. Since sound has a significantly large wavelength as compared to light, it can diffract through a much larger opening than light can. The relationship between the angle of diffraction (θ), wavelength (λ), and opening size (d) is as follows:

$$\lambda = d \sin \theta$$

Note that, for a given wavelength, the opening size and the angle of diffraction are inversely related. This can also be seen in the diagrams above. As the opening size gets smaller, the pattern spreads out to a larger angle (produces a greater amount of diffraction).

With light, for example, passing a laser through a small opening produces an interesting effect. A pattern of alternating bright and dark bands is produced when projected onto a screen, as in the image below:

We should note that there is a wider bright spot in the center, and then narrower alternating bright and dark bands on either side. This is a characteristic of a *single slit diffraction pattern.*

What would happen if waves were passed through two openings instead of one? Using a similar example with water waves, passing the waves through two narrow openings relatively close together produces an effect like:

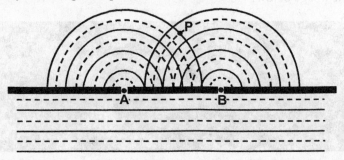

If the solid lines represent crests of the water waves and the dashed lines represent the troughs, there is an overlap of the diffracted waves from openings A and B. Using the principle of *superposition*, it can be determined whether the diffracted waves from the two openings produce *constructive interference* or *destructive interference.* For example, at point P above, a crest from the diffracted wave from opening A is combined with a trough from the diffracted wave from opening B. Therefore, as a result, there is destructive interference at point P, and there is actually no wave energy there. In locations where there are two crests, or two troughs, then there would be constructive interference.

Thomas Young

Thomas Young was an English physician and physicist who observed the interference of light. He noted that when light from a single source is separated into two beams and these beams are recombined, the combined beams produce a pattern of bright and dark fringes (Young's double-slit experiment). He also explained Young's modulus - the relationship between stress and strain; and partly deciphered the Rosetta Stone.

Born: June 13, 1773; England
Died: May 10, 1829; England

In the early 1800s, Thomas Young performed what is now called Young's Double Slit Experiment. In this demonstration, Young passed **coherent light** through two narrow vertical slits and observed the pattern formed. Coherent light is light that produces waves with a constant phase relationship. Many of the early experiments used **monochromatic** light (one color, or frequency). Today, a monochromatic, coherent light source that can be used for these demonstrations is a laser.

The pattern that Young observed was an alternating pattern of bright and dark bands, similar to the image below:

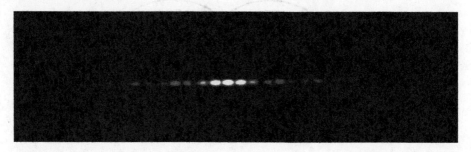

Here we can see that there is a pattern of evenly spaced, equally sized alternating bright and dark bands. This is a characteristic of a *double slit diffraction pattern.*

What would happen if there were more than two slits? What if there were hundreds, or even thousands of slits? An optical device consisting of a large number of narrow parallel slits that allows light to pass through is called a transmission **diffraction grating**.

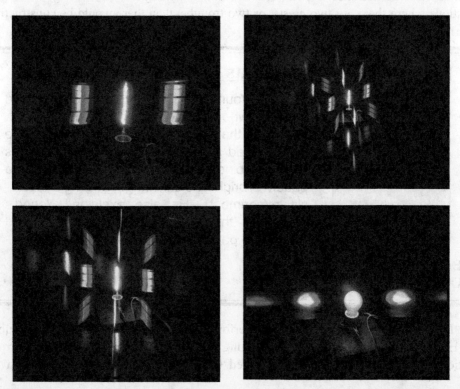

A diffraction grating produces a pattern similar to a double-slit diffraction pattern for monochromatic light. When polychromatic light (many colors) is passed through a diffraction grating, however, the pattern shows the light separated into its component colors for all of the bright bands (called maxima) except for the bright band in the middle (called the central maximum). The central maximum is the original color of the undiffracted light.

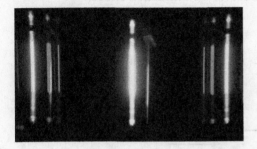 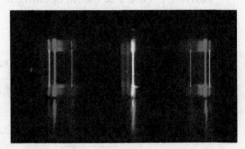

Any polychromatic light has its own characteristic spectrum when passed through a diffraction grating. Using the spectrum pattern to identify the elements in a gas emission tube, for example, is one practical application of diffraction gratings.

Unit 9
Practice Questions

1. A wave of constant wavelength diffracts as it passes through an opening in a barrier. As the size of the opening is increased, the diffraction effects

    (1) decrease   (2) increase   (3) remain the same

2. The diagram below represents two pulses approaching each other from opposite directions in the same medium.

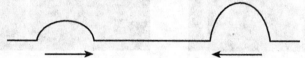

    Which diagram best represents the medium after the pulses have passed through each other?

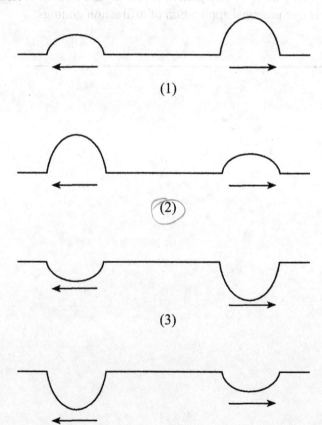

(1)

(2)

(3)

(4)

DIFFRACTION AND INTERFERENCE

3. A beam of monochromatic light approaches a barrier having four openings, A, B, C, and D, of different sizes as shown below.

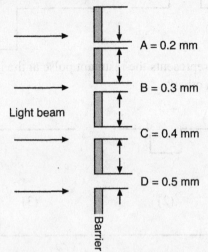

Light beam

A = 0.2 mm

B = 0.3 mm

C = 0.4 mm

D = 0.5 mm

Barrier

Which opening will cause the greatest diffraction?

(1) A   (2) B   (3) C   (4) D

---

4. Two waves having the same frequency and amplitude are traveling in the same medium. Maximum constructive interference occurs at points where the phase difference between the two superposed waves is

(1) 0°   (2) 90°   (3) 180°   (4) 270°

5. Parallel wave fronts incident on an opening in a barrier are diffracted. For which combination of wavelength and size of opening will diffraction effect be greatest?

(1) short wavelength and narrow opening
(2) short wavelength and wide opening
(3) long wavelength and narrow opening
(4) long wavelength and wide opening

6. The diagram below represents two pulses approaching each other.

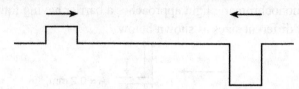

Which diagram best represents the resultant pulse at the instant the pulses are passing through each other?

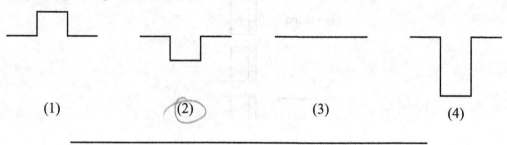

(1) (2) (3) (4)

7. The diagram below shows two pulses of equal amplitude, A, approaching point P along a uniform string.

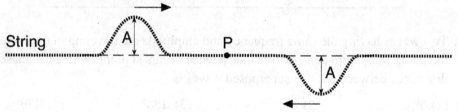

When the two pulses meet at P, the vertical displacement of the string at P will be

(1) A (2) 2A (3) 0 (4) $\frac{A}{2}$

8. A wave is diffracted as it passes through an opening in a barrier. The amount of diffraction that the wave undergoes depends on both the

(1) amplitude and frequency of the incident wave
(2) wavelength and speed of the incident wave
(3) wavelength of the incident wave and the size of the opening
(4) amplitude of the incident wave and the size of the opening

9. The diagram below represents two waves of equal amplitude and frequency approaching point P as they move through the same medium.

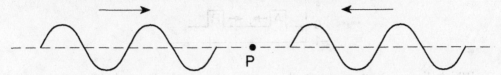

As the two waves pass through each other, the medium at point P will

(1) vibrate up and down   (3) vibrate into and out of the page
(2) vibrate left and right   (4) remains stationary

---

10. Two pulses, A and B, travel toward each other along the same rope, as shown below.

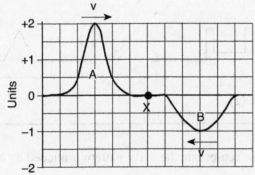

When the centers of the two pulses meet at point X, the amplitude at the center of the resultant pulse will be

(1) +1 unit   (2) +2 units   (3) 0   (4) -1 unit

---

11. The diagram below shows two pulses, A and B, approaching each other in a uniform medium.

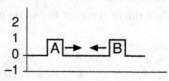

Which diagram best represents the superposition of the two pulses?

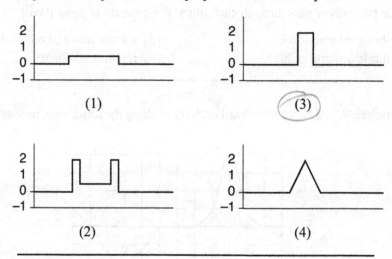

(1)   (3)

(2)   (4)

12. The diagram below represents the wave pattern produced by two sources located at point A and B.

Wave fronts

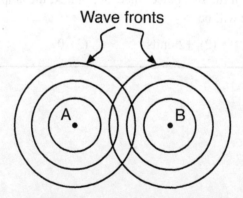

Which phenomenon occurs at the intersections of the circular wave fronts?

(1) diffraction   (2) interference   (3) refraction   (4) reflection

1. An electromagnetic AM-band radio wave could have a wavelength of

   (1) 0.005 m   (3) 500 m
   (2) 5 m   (4) 5 000 000 m

2. A car's horn is producing a sound wave having a constant frequency of 350 hertz. If the car moves toward a stationary observer at constant speed, the frequency of the car's horn detected by this observer may be

   (1) 320 Hz   (2) 330 Hz   (3) 350 Hz   (4) 380 Hz

3. At an outdoor physics demonstration, a delay of 0.50 second was observed between the time sound waves left a loudspeaker and the time these sound waves reached a student through the air. If the air is at STP, how far was the student from the speaker?

   (1) 1.5×10^{-3} m   (2) 1.7×10^2 m   (3) 6.6×10^2 m   (4) 1.5×10^8 m

4. Which type of wave requires a material medium through which to travel?

   (1) electromagnetic   (3) sound
   (2) infrared   (4) radio

5. As a transverse wave travels through a medium, the individual particles of the medium move
   (1) perpendicular to the direction of wave travel
   (2) parallel to the direction of wave travel
   (3) in circles
   (4) in ellipses

6. A 512-hertz sound wave travels 100. meters to an observer through air at STP. What is the wavelength of this sound wave?

   (1) 0.195 m   (2) 0.646 m   (3) 1.55 m   (4) 5.12 m

7. Which pair of terms best describes light waves traveling from the Sun to Earth?

   (1) electromagnetic and transverse
   (2) electromagnetic and longitudinal
   (3) mechanical and transverse
   (4) mechanical and longitudinal

8. Compared to the period of a wave of red light the period of a wave of green light is

 (1) less   (2) greater   (3) the same

9. Electrons oscillating with a frequency of 2.0×10^{10} hertz produce electromagnetic waves. These waves would be classified as

 (1) infrared   (2) visible   (3) microwave   (4) x ray

10. An electric bell connected to a battery is sealed inside a large jar. What happens as the air is removed from the jar?

 (1) The electric circuit stops working because electromagnetic radiation can *not* travel through a vacuum.
 (2) The bell's pitch decreases because the frequency of the sound waves is lower in a vacuum than in air.
 (3) The bell's loudness increases because of decreased air resistance.
 (4) The bell's loudness decreases because sound waves can *not* travel through a vacuum.

11. A student strikes the top rope of a volleyball net, sending a single vibratory disturbance along the length of the net, as shown in the diagram below.

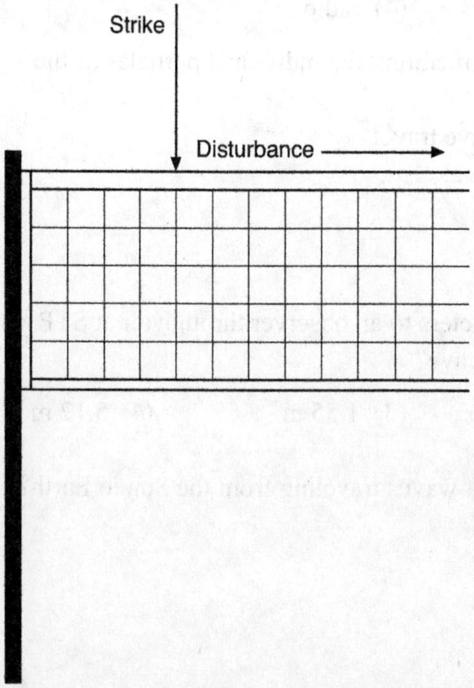

Strike

Disturbance ⟶

This disturbance is best described as

 (1) a pulse
 (2) a periodic wave
 (3) a longitudinal wave
 (4) an electromagnetic wave

12. How much time does it take light from a flash camera to reach a subject 6.0 meters across a room?

(1) 5.0×10^{-9} s   (2) 2.0×10^{-8} s   (3) 5.0×10^{-8} s   (3) 2.0×10^{-7} s

13. The diagram below shows a periodic wave.

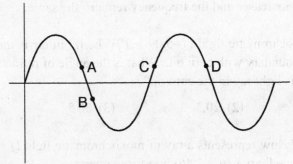

Which points are in phase with each other?

(1) A and C   (2) A and D   (3) B and C   (4) C and D

---

14. An electric guitar is generating a sound of constant frequency. An increase in which sound wave characteristic would result in an increase in loudness?

(1) speed   (3) wavelength
(2) period   (4) amplitude

15. In a vacuum, light with a frequency of 5.0×10^{14} hertz has a wavelength of

(1) 6.0×10^{-21} m   (3) 1.7×10^{6} m
(2) 6.0×10^{-7} m   (4) 1.5×10^{23} m

16. When a light wave enters a new medium and is refracted, there must be a change in the light wave's

(1) color   (3) period
(2) frequency   (4) speed

17. The speed of light in a piece of plastic is 2.00×10^{8} meters per second. What is the absolute index of refraction of this plastic?

(1) 1.00   (2) 0.670   (3) 1.33   (4) 1.50

18. What happens to the speed and frequency of a light ray when it passes from air into water?

(1) The speed decreases and the frequency increases.
(2) The speed decreases and the frequency remains the same.
(3) The speed increases and the frequency increases.
(4) The speed increases and the frequency remains the same.

19. A ray of monochromatic light ($f=5.09 \times 10^{14}$ hertz) in air is incident at an angle of 30.° on a boundary with corn oil. What is the angle of refraction, to the nearest degree, for this light ray in the corn oil?

(1) 6°   (2) 20.°   (3) 30.°   (4) 47°

20. The diagram below represents a ray of monochromatic light ($f = 5.09 \times 10^{14}$ Hz) passing from medium X (n = 1.46) into fused quartz.

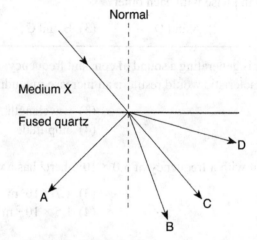

Which path will the ray follow in the quartz?

(1) A   (2) B   (3) C   (4) D

21. Two waves having the same amplitude and frequency are traveling in the same medium. Maximum destructive interference will occur when the phase difference between the waves is

(1) 0°   (2) 90°   (3) 180°   (4) 270°

22. Which wave phenomenon occurs when vibrations in one object cause vibrations in a second object?

(1) reflection   (2) resonance   (3) intensity   (4) tuning

23. The diagram below represents shallow water waves of constant wavelength passing through two small openings, *A* and *B*, in a barrier.

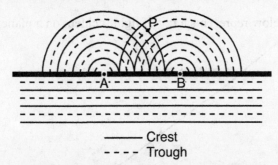

———— Crest
- - - - Trough

Which statement best describes the interference at point P?

(1) It is constructive, and causes a longer wavelength.
(2) It is constructive, and causes an increase in amplitude.
(3) It is destructive, and causes a shorter wavelength.
(4) It is destructive, and causes a decrease in amplitude.

24. The superposition of two waves traveling in the same medium produces a standing wave pattern if the two waves have

(1) the same frequency, the same amplitude, and travel in the same direction
(2) the same frequency, the same amplitude, and travel in the opposite directions
(3) the same frequency, different amplitudes, and travel in the same direction
(4) the same frequency, different amplitudes, and travel in opposite directions

25. The diagram to the right shows two pulses traveling toward each other in a uniform medium.

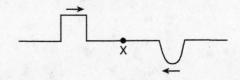

Which diagram best represents the medium when the pulses meet at point *X*?

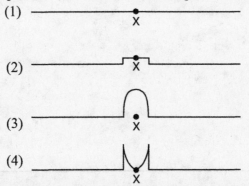

Part B2

1. The diagram below represents a ray of light incident on a plane mirror.

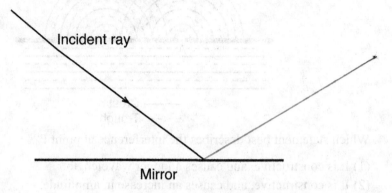

Incident ray

Mirror

Using a protractor and straightedge, on the diagram above, construct the reflected ray for the incident ray shown.

2. A beam of monochromatic light has a wavelength of 5.89×10^{-7} meter in air. Calculate the wavelength of this light in diamond. (Show all work, including the equation and substitution with units.)

$$\frac{n_2}{n_1} = \frac{\lambda_1}{\lambda_2}$$

$$\frac{2.42}{1.00} = \frac{5.89 \times 10^{-7}}{\lambda_2}$$

$$\boxed{\lambda_2 = 2.43 \times 10^{-7} \, m}$$

3. The diagram below represents a transverse wave moving on a uniform rope with point A labeled as shown. On the diagram mark an **X** at the point on the wave that is 180° out of phase with point A.

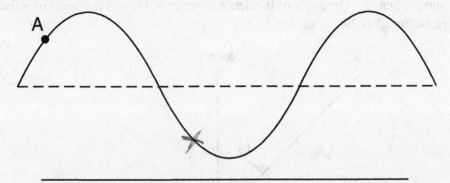

4. The graph below represents the relationship between wavelength and frequency of waves created by two students shaking the ends of a loose spring.

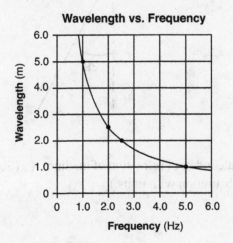

Calculate the speed of the waves generated in the spring. (Show all work, including the equation and subsitution with units.)

$$V = \lambda f$$
$$V = (5m)(1\,Hz)$$
$$\boxed{V = 5\,m/s}$$

Base your answers to questions 5 and 6 on the information and diagram below.

A ray of monochromatic light of frequency 5.09×10^{14} hertz is traveling from water into medium X. The angle of incidence in water is 45° and the angle of refraction in medium X is 29°, as shown.

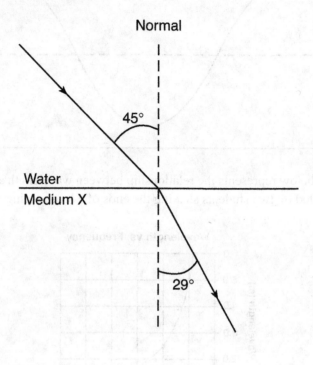

5. Calculate the absolute index of refraction of medium X. (Show all work, including the equation and substitution with units.)

$$n_1 \sin\Theta_1 = n_2 \sin\Theta_2$$
$$1.33 \sin 45° = n_2 \sin 29°$$
$$\boxed{n_2 = 1.94}$$

6. Medium X is most likely what material? _____Zircon_____

7. An FM radio station broadcasts its signal at a frequency of 9.15×10^7 hertz. Determine the wavelength of the signal in air.

$$v = \lambda f$$
$$3 \times 10^8 = (9.15 \times 10^7 \, Hz) \lambda$$
$$\lambda = 3.28 \, m$$

_____3.28_____ m

8. A ray of monchromatic light with a frequency of 5.09×10^{14} hertz is transmitted through four different media, listed below.

> **A.** corn oil
> **B.** ethyl alcohol
> **C.** flint glass
> **D.** water

Rank the four media from the one through which the light travels at the slowest speed to the one through which the light travels at the fastest speed. (Use the letters in front of each medium to indicate your answer.)

___D___   ___B___   ___A___   ___C___

9. The diagram below represents a transverse wave moving along a string.

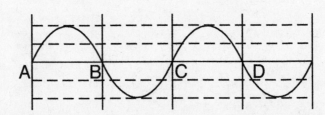

On the diagram below, draw a transverse wave that would produce complete destructive interference when superimposed with the original wave.

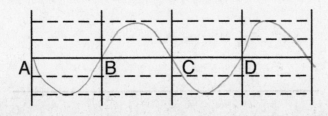

Base your answers to question 10 through 12 on the information and diagram below.

A light ray with a frequency of 5.09×10^{14} hertz traveling in air is incident at an angle of 40.° on an air-water interface as shown. At the interface, part of the ray is refracted as it enters the water and part of the ray is reflected from the interface.

10. Calculate the angle of refraction of the light ray as it enters the water. (Show all work in the space provided below, including the equation and substitution. with units.)

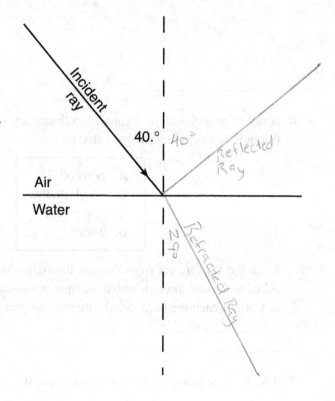

11. On the diagram, using a protractor and straightedge, draw the refracted ray. Label this ray "Refracted Ray".

12. On the diagram, using a protractor and straightedge, draw the reflected ray. Label this ray "Reflected Ray".

$1 \sin 40° = 1.33 \sin \theta$

$\theta = 29°$

13. A student plucks a guitar string and the vibrations produce a sound wave with a frequency of 650 hertz. Calculate the wavelength of the sound wave in air at STP. (Show all work, including the equation and substitution with units.)

$$v = \lambda f$$
$$(3 \times 10^8 \text{ m/s}) = \lambda (650 \text{ Hz})$$
$$\boxed{\lambda = 4.62 \times 10^5}$$

Base your answers to questions 14-16 on the information and diagram below.

Three waves, A, B, and C, travel 12 meters in 2.0 seconds through the same medium as shown in the diagram below.

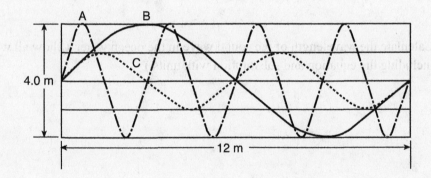

14. What is the amplitude of wave C? ___1___ m

15. What is the period of wave A? ___0.5___ s

16. What is the speed of wave B? ___6___ m/s

Part C

Base your answers to question 1 through 3 on the information below.

A stationary research ship uses sonar to send a 1.18×10^3-hertz sound wave down through the ocean water. The reflected sound wave from the flat ocean bottom 324 meters below the ship is detected 0.425 second after it was sent from the ship.

1. Calculate the speed of the sound wave in the ocean water. (Show all work, including the equation and substitution with units.)

2. Calculate the wavelength of the sound wave in the ocean water. (Show all work, including the equation and substitution with units.)

3. Determine the period of the sound wave in the ocean water.

_____ s

Base your answers to question 4 through 6 on the information and diagram below.

A ray of monochromatic light having a frequency of 5.09×10^{14} hertz is incident on an interface of air and corn oil at an angle of 35° as shown. The ray is transmitted through parallel layers of corn oil and glycerol and is then reflected from the surface of a plane mirror, located below and parallel to the glycerol layer. The ray then emerges from the corn oil back into the air at point P.

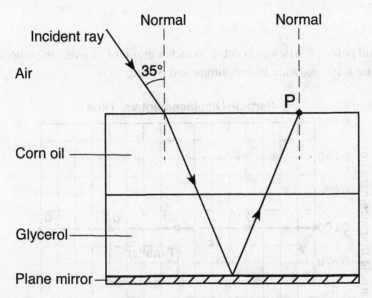

4. On the diagram above, use a protractor and straightedge to construct the refracted ray representing the light emerging at point P into air.

5. Calculate the angle of refraction of the light ray as it enters the corn oil from air. (Show all work, including the equation and substitution with units.)

6. Explain why the ray does not bend at the corn oil-glycerol interface.

_____

_____

Base your answers to question 7 through 9 on the information below.

A periodic wave traveling in a uniform medium has a wavelength of 0.080 meter, an amplitude of 0.040 meter, and a frequency of 5.0 hertz.

7. Determine the period of the wave.

_____ s

8. On the grid below, starting at point *A*, sketch a graph of *at least one* complete cycle of the wave showing its amplitude and period.

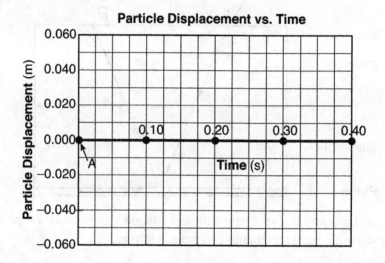

Particle Displacement vs. Time

9. Calculate the speed of the wave. (Show all work, including the equation and substitution with units.)

Base your answers to question 10 and 11 on the information and diagram below.

Sunlight is composed of various intensities of all frequencies of visible light. The graph represents the relationship between light intensity and frequency.

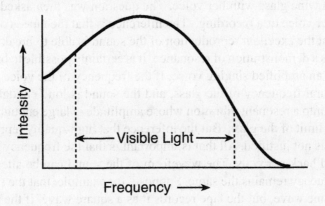

10. Based on the graph, which color of visible light has the lowest intensity?

_____

11. It has been suggested that fire trucks be painted yellow-green instead of red. Using information from the graph, explain the advantage of using yellow-green paint.

_____

_____

_____

Shattering Glass

An old television commercial for audio recording tape showed a singer breaking a wine glass with her voice. The question was then asked if this was actually her voice or a recording. The inference is that the tape is of such high quality that the excellent reproduction of the sound is able to break glass.

This is a demonstration of resonance. It is certainly possible to break a wine glass with an amplified singing voice. If the frequency of the voice is the same as the natural frequency of the glass, and the sound is loud enough, the glass can be set into a resonant vibration whose amplitude is large enough to surpass the elastic limit of the glass. But the inference that high-quality reproduction is necessary is not justified. All that is important is that the frequency is recorded and played back correctly. The waveform of the sound can be altered as long as the frequency remains the same. Suppose, for example, that the singer sings a perfect sine wave, but the tape records it as a square wave. If the tape player plays the sound back at the right speed, the glass will still receive energy at the resonance frequency and will be set into vibration leading to breakage, even though the tape reproduction was terrible. Thus, this phenomenon does not require high-quality reproduction and, thus, does not demonstrate the quality of the recording tape. What it does demonstrate is the quality of the tape player, in that it played back the tape at an accurate speed!

12. List two properties that a singer's voice must have in order to shatter a glass.

_____

_____

13. Explain why the glass would not break if the tape player did not play back at an accurate speed.

_____

_____

_____

_____

PART III:
ELECTRICITY AND MAGNETISM

UNIT 10
Static Electricity and Fields

Static electricity refers to charges that are at rest. In this unit, the properties of static electricity, electric forces, and electric fields will be explored. What exactly makes an object charged, how the charges interact with each other, and the relationship between atomic structure and charge will also be discussed.

Atomic Structure and charge

The basic structure of the atom consists of subatomic particles called *protons, neutrons,* and *electrons*. It will be discussed in future units that these are not the smallest subatomic particles, but for a discussion of electric properties of matter, this is sufficient. The proton is a positively charged particle, the electron is negatively charged, and the neutron is neutral.

If an atom has an equal number of protons and electrons, that atom is said to be **electrically neutral**. If an atom has more electrons than it has protons, it is **negative**, and if it has more protons than electrons, it is **positive**. A charged atom is typically called an *ion*.

On a larger scale, if an entire object consisting of a very large number of atoms has more protons than electrons, it is likewise positive. If the entire object has more electrons than protons, it is negative.

Acquisition of charge

An object can become charged if it acquires an imbalance among protons and electrons. This is the result of the *transfer of electrons* to or from the object. The total number of protons will not change - if it did change, this would require a nuclear process!

In order to transfer electrons to or from an object, work must be done. Recall from earlier units that work is the product of force and displacement. Often times, the force that is responsible for this charge transfer is from the friction of two objects rubbing together.

There are many situations where a person has walked across a carpeted floor, perhaps dragging his feet a bit, and when he reaches out to a doorknob... zap! This zap is the *discharge* of electrons on that person who became charged while dragging his feet. The friction between the person's feet and the carpet resulted in a large number of electrons transferring to the person and then, when touching the doorknob, they traveled to the doorknob to make the person *neutral* again.

A summary of charged objects is as follows:

- A neutral object has an equal number of protons and electrons
- A positive object has a deficiency (lack) of electrons
- A negative object has an excess of electrons
- An object loses its excess charge and becomes neutral when grounded

The tendency of an object to lose electrons and become positive, or for an object to gain electrons and become negative is based on its elemental and molecular composition. How tightly bound the electrons are to the atoms and molecules will determine whether it will gain or lose electrons. The *triboelectric series* is a list of many common materials and their tendencies to become positive or negative.

TRIBOELECTRIC SERIES
More Positive
Acrylic, Lucite
Leather
Rabbit's Fur
Glass
Quartz
Mica
Nylon
Cat's Fur
Silk
Amber
Polystyrene
Rubber
Plastic wrap
Polyethylene ("Scotch tape")
PVC
Ebonite
More Negative

If two objects from this list are rubbed together, the material higher on the list will become positive and the lower material will become negative. There are two very common combinations used in the Physics classroom or laboratory:

- **Rubber rod rubbed with animal fur**: The rubber rod will become negatively charged by gaining electrons and the animal fur will become positive by losing electrons (to the rubber rod).
- **Glass rod rubbed with silk:** The glass rod will become positively charged by losing electrons and the silk will become negative by gaining electrons (from the glass rod).

Conductors and Insulators

Some of the materials listed in the above table are **conductors**, which means that they have the ability to allow charges to move relatively freely. Metals, such as gold, silver, and platinum are good conductors. Materials that do not allow charges to move freely are categorized as **insulators**. Some examples of insulators include air, glass, wood, plastic, and rubber.

Conservation of Charge

The net charge in a closed system remains constant. Electrons cannot be created or destroyed, but they can be transferred from one object to another. In a closed system with two objects, the number of electrons that one object loses is equal to the number of electrons that the other object gains.

Measurement of Charge

How much charge an object has acquired (q) can be measured by the number of electrons, or **elementary charges (e)**, it has gained or lost. Typically, however, objects will gain or lose a very large number of elementary charges. A representation of a very large number of elementary charges is called the **Coulomb (C)**, named for Charles-Augustin de Coulomb. From the reference tables:

$$1 \text{ Coulomb} = 6.25 \times 10^{18} \text{ elementary charges}$$

Since the Coulomb represents a very large number of elementary charges, the charge of one proton or electron can be expressed in terms of Coulombs:

$$1 \text{ elementary charge} = 1.60 \times 10^{-19} \text{ Coulombs}$$

The formula for determining the charge on an object, in Coulombs, if the number of elementary charges is known is:

$$q = n \cdot e$$

where:
q = charge of the object, measured in Coulombs
n = number of elementary charges
e = the charge of 1 elementary charge (1.60×10^{-19} C)

Charles-Augustin de Coulomb

Charles-Augustin de Coulomb was a French physicist, mathematician, and engineer most known for his quantitative determination of Coulomb's Law. Using a torsion balance, Coulomb made measurements of the forces of attraction and repulsion of electrified bodies. In France, he is considered one of the pioneers of experimental science. The SI unit of charge (the Coulomb, C) is named after him.

Born: June 14, 1736; France
Died: August 23, 1806; France

Served in the French army as a military engineer in charge of building Fort Bourbon in the West Indies

The Millikan Experiment and the elementary charge

One of the more famous experiments in physics with respect to electricity was the oil drop experiment performed by Robert Millikan and reported in 1913. In this experiment, Millikan measured the forces necessary to suspend charged drops of oil in a uniform electric field. He found that no drop had a charge less than 1.60×10^{-19} C. The charges on each of the other drops were multiples of this value. This finding demonstrated that there is a fundamental unit of charge, called the elementary unit of charge: 1.60×10^{-19} C.

Robert Millikan

Robert Millikan was an American experimental physicist most known for experimentally determining the charge of an electron with the oil drop experiment. Millikan also was skeptical of Einstein's particle theory of light for the photoelectric effect and conducted research to disprove it, only to find data that agreed with the particle theory.

Born: March 22, 1868; USA
Died: December 19, 1953; USA

Colleges Attended: Oberlin College, Columbia University
Nobel Prize in Physics, 1923

Solved Example Problems

1. A sphere has a negative charge of 6.4×10^{-7} Coulomb. Approximately how many electrons must be removed to make the sphere electrically neutral?

$$q = n \cdot e$$
$$6.4 \times 10^{-7}\,C = n \cdot (1.60 \times 10^{-19}\,C)$$
$$n = 4.0 \times 10^{12}\ \text{elementary charges}$$

2. How many Coulombs of charge do 3 elementary charges have?

$$q = n \cdot e$$
$$q = 3 \cdot (1.60 \times 10^{-19}\,C)$$
$$q = 4.80 \times 10^{-19}\,C$$

3. The diagram below shows two metal spheres, A and B, on insulated stands. Sphere A possesses a net charge of -3.0×10^{-6} C, and sphere B possesses a net charge of -6.0×10^{-6} C. If the spheres are brought into contact with each other and then separated, what will be the charge on each sphere?

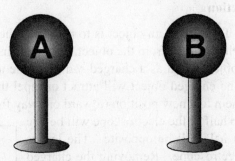

The charges will equalize on the two objects and each will have the same magnitude of charge. This is found most easily by finding the mathematical average of the two objects' charges.

$$\frac{(-3.0 \times 10^{-6}\,C) + (-6.0 \times 10^{-6}\,C)}{2} = -4.5 \times 10^{-6}\,C$$

This is equivalent to -4.5μC

Detecting and Visualizing Charge

A device used to detect the presence and show relative amounts of charge is an **electroscope**. An electroscope is often made with light metal foil "leaves" attached to a conducting rod and electrically insulated from the surroundings. Some examples of electroscopes are below:

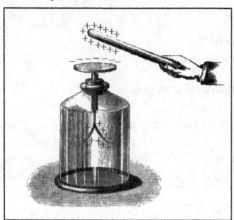
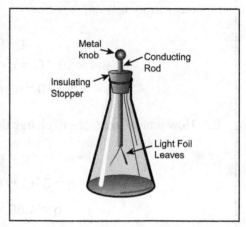

When an electroscope is neutral, the leaves will hang straight down. If the leaves are charged, they will repel each other because they will have the same charge. The leaves of an electroscope cannot carry opposite charges because they are connected to each other.

Methods of Using an Electroscope

There are three main ways to use an electroscope. These methods are: **polarization, conduction, and induction**.

1. Polarization: To polarize an object is to rearrange the charges within it, but there is no transfer of charge to or from the object. In polarizing an electroscope, one must bring a charged object (such as a charged rod) near the top of the electroscope without touching it. The charged object will attract or repel the electrons within the electroscope moving them to a new position (toward or away from the rod, depending on the charge). The top half of the electroscope will be one charge and the bottom half will be opposite. The leaves will separate on the electroscope. Removing the charged rod will allow the charges within the electroscope to redistribute evenly and the leaves will drop.

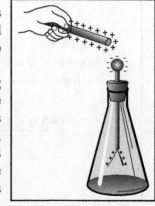

In the diagram on the right, the electroscope is being polarized by a positively charged rod. Bringing the positive rod near the top of the electroscope results in the electrons within the electroscope being attracted toward the top, leaving the bottom part positive and the top negative. This rearrangement of the charges will cause the leaves at the bottom to spread apart. This is only temporary; the leaves will return back after the rod is removed.

2. Conduction, or contact: Charging an electroscope by conduction requires physical contact with the electroscope. In this method, charge is transferred to or from the electroscope. When charging an object by **conduction**, the object (or electroscope) will have the **same charge** as the object used to charge it.

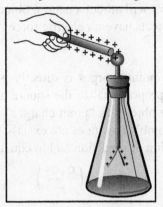

In the example above, a positive rod is touched directly to the top of the electroscope. Since the rod is positive, it needs to acquire electrons to neutralize it. The rod will remove some electrons from the electroscope. It will not be able to remove enough electrons to neutralize the rod, so it will be still positive (although not as much), and the entire electroscope will become positive because of the loss of electrons to the rod. The leaves will spread apart. When the rod is removed, the entire electroscope will be positively charged and the leaves will remain apart until the electroscope is **grounded** to allow it to become neutral. A ground is a source or receiver of charge. An electroscope can be safely grounded by touching it with a finger.

Induction: Charging an object, or electroscope, by induction is a three step process.

 i. *Polarize the electroscope* by bringing a charged object near the top, without touching it.

 ii. *Ground the electroscope* with the charged object still near it.

 iii. *Remove the ground first, then the charged object.*

When charging an object by **induction**, the object (or electroscope) will have the **opposite charge** as compared to the object used to charge it. An example of the steps necessary to charge an electroscope negatively (using a positive rod) using induction is in the diagram on the right:

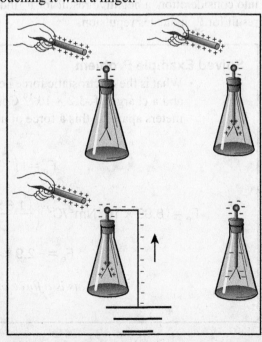

Interactions among charged objects

Charged objects will interact with each other according to the following rules:
- like charges repel
- opposite charges attract
- a charged object will attract a neutral object
- two neutral objects have no electric force between them

Coulomb's Law

The force between two point charges is directly proportional to the product of their charges and inversely proportional to the square of the distance between them. Coulomb's Law refers to the objects as "point charges" and if the charges objects are relatively large in size, Coulomb's Law does not exactly apply. This force is called the **electrostatic force** of attraction or repulsion and in equation form:

$$F_e = k\left(\frac{q_1 q_2}{r^2}\right)$$

where:

F_e = electrostatic force, measured in Newtons
k = electrostatic constant (8.99×10^9 Nm^2/C^2)
q_1, q_2 = charge on each object, measured in Coulombs
r = distance between objects, measured in meters

It should be noted that the inverse square structure of this formula closely resembles Newton's Law of Universal Gravitation ($F_g = G\left(\frac{m_1 m_2}{r^2}\right)$). A major difference, however, is that the electrostatic force may be attractive or repulsive, as compared to the gravitational force, which is only attractive. With the signs of the charges taken into consideration, a negative result for F_e indicates a force of attraction, and a positive result for F_e indicates repulsion.

Solved Example Problem

- What is the electrostatic force between a charge of 1.6×10^{-19} Coulomb and a charge of -3.2×10^{-19} Coulomb when they are a distance of 4.0 meters apart? Is this a force of attraction or repulsion?

$$F_e = k\left(\frac{q_1 q_2}{r^2}\right)$$

$$F_e = (8.99 \times 10^9 \, Nm^2/C^2)\left(\frac{(1.6 \times 10^{-19}C)(-3.2 \times 10^{-19}C)}{(4.0m)^2}\right)$$

$$F_e = -2.9 \times 10^{-29} \, N$$

This is a force of attraction.

Electric Fields

An **electric field** is a region in which electric forces act on charges. It is determined by identifying the magnitude and direction of the force exerted on a positive . The strength of an electric field, or **electric field intensity** is a vector quantity, having magnitude and direction.

The magnitude of the electric field intensity is determined by the force per unit charge. In equation form:

$$E = \frac{F_e}{q}$$

where:
E = electric field strength, measured in N/C
F_e = force exerted on the test charge, measured in Newtons
q = charge, measured in Coulombs

The direction of an electric field is determined by the direction of the force on a very small positive test charge. If an object has a negative charge, the positive test charge will be attracted to it, so the direction of the electric field near this negative object is toward it. Likewise, if an object is positive, a test charge will be repelled, indicating that the direction of the electric field is away from it.

There are many similarities between electric fields and gravitational fields. The strength of both is represented pictorially by the relative number and spacing of the field lines. Unlike the gravitational field, which only is directed toward an object, the electric field may point toward or away, depending on the sign of the object's charge. The gravitational field is force per unit mass, and the electric field is force per unit charge.

The equation above for the strength of an electric field can be combined with Coulomb's Law with the test charge as one of the charged objects.

$$E = \frac{F_e}{q} \text{ and } F_e = k\left(\frac{q_1 q_2}{r^2}\right)$$

$$E = \frac{k\frac{q_1 q_2}{r^2}}{q}$$

$$E = k\left(\frac{q}{r^2}\right)$$

The formula above represents a way to calculate the strength of an electric field any distance away from a charged object. The electric field intensity varies inversely with the square of the distance from that charge.

There are some specific rules to consider when drawing electric field lines. The rules are:
- Field lines represent the force exerted on a small positive imaginary test charge.
- Field lines point away from positive objects.
- Field lines point toward negative objects.
- Field lines are perpendicular to the surface of an object.
- Field lines never cross.

It is important to know the pictorial representation of the electric field in the vicinity of some specific charges and charge combinations. They are as follows:
1. a positive point charge
2. a negative point charge
3. two positive point charges
4. two negative point charges
5. one positive and one negative point charge
6. oppositely-charged parallel plates

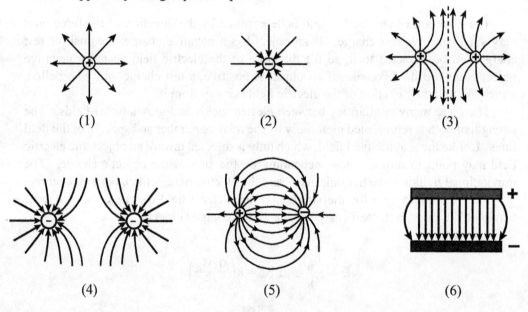

(1)   (2)   (3)

(4)   (5)   (6)

Since the relative strength of the electric field is represented by the spacing of the field line vectors, it can be determined that the field is strongest near a single point charge and weakens greatly with increased distance. With two oppositely charged objects, the field is strongest between the objects. With two similarly charged objects, the field is weakest between the objects. With parallel plates, the field is of uniform strength between the plates, as shown by the parallel field lines. At the edges of the plates, however, the strength varies slightly. If the distance between the plates is small as compared to their surface area, this effect is minimal.

Solved Example Problems

1. What is the intensity of an electric field where an electron experiences a force of 3.2×10^{-15}N?

$$E = \frac{F_e}{q}$$

$$E = \frac{3.2 \times 10^{-15} N}{1.6 \times 10^{-19} C}$$

$$E = 2.0 \times 10^4 N/C$$

2. If the intensity at a point in an electric field is 8.0 newtons per Coulomb, then what is the force on a charge of 2.0 Coulombs at that point in the field?

$$E = \frac{F_e}{q}$$

$$F_e = qE$$

$$F_e = (2.0\,C)(8.0\,N/C)$$

$$F_e = 16 \text{ N}$$

Electric Potential and Potential Difference

Imagine that you wanted to move a positive test charge from an infinite distance to a point closer to a positively charged object. Because these charges repel, it requires work in order to do this. The electric potential at any point in an electric field is the work needed to move this test charge from infinity to that location. Much like work is needed to move an object against a gravitational field, work is needed to move a charged object against an electric field, and the electric potential energy increases. If the electric field does the work (i.e. the charge is moving with the electric field) the potential energy of the charge decreases.

Consider two points in an electric field in the vicinity of a positive charge, A and B. If B is closer to the object than A, moving a positive test charge from A to B requires work and results in an increase in the potential energy of the test charge. This change is called the potential difference (Voltage) between the two points.

Potential difference is measured with the **Volt (V)**, named after Alessandro Volta. The Volt is defined as the potential difference between two points whose positions in the field are such that one joule of work is required to transfer one Coulomb of charge from one position to another. The formula for determining potential difference is:

$$V = \frac{W}{q}$$

where:
V = potential difference, measured in Volts (V)
W = Work, measured in Joules (J)
q = charge, measured in Coulombs (C)

Therefore, it is shown from this formula that $1V = 1$ J/C.

Alessandro Volta

Alessandro Volta was an Italian physicist most known for his invention of the Voltaic Pile, an early battery. He is also credited with improving the electrophorus, discovering methane gas, and researching electrical capacitance. The SI unit of electric potential (the Volt, V) is named after him.

Born: February 18, 1745; Italy
Died: March 5, 1827; Italy

The **electron-volt (eV)** is a measurement of energy. It is the work required to move one elementary charge through a potential difference of one volt. Since the charge on an electron is 1.60×10^{-19} C, and one Volt is one Joule per Coulomb, then:

$$V = \frac{W}{q}$$

$$1\,{}^{J}\!/_{C} = \frac{W}{1.60 \times 10^{-19}\,C}$$

$$W = 1.60 \times 10^{-19}\,J = 1eV$$

Solved Example Problem

• An elementary charge is accelerated by a potential difference of 12.0 volts. The total energy acquired by the charge in electron-volts and Joules is:

$$V = \frac{W}{q}$$

$$12.0\,V = \frac{W}{1.60 \times 10^{-19}\,C}$$

$$W = 1.92 \times 10^{-18}\,J$$

$$1.92 \times 10^{-18}\,J\left(\frac{1eV}{1.60 \times 10^{-19}\,J}\right) = 12.\,eV$$

Electric Field Strength in terms of Potential

Electric Field strength can be expressed in terms of potential difference. Combining the formula for electric field strength with the formula for potential difference yields:

$$E = \frac{F_e}{q} \text{ and } V = \frac{W}{q}$$

$$(\text{where } W = F_e d) \gg V = \frac{F_e d}{q}$$

solve the above for F_e: $F_e = \frac{v_q}{d}$ and combine with $E = \frac{F_e}{q}$

$$E = \frac{\left(\frac{v_q}{d}\right)}{q}$$

$$E = \frac{V}{d}$$

This allows the electric field strength to be calculated in terms of the potential difference. This is especially useful for calculating the electric field strength between parallel plates if the Voltage (potential difference) and distance between the plates are known. Since it is known that the unit for electric field strength is N/C and, with the formula above, the unit would be V/m, their equivalence can be shown as:

$$\frac{V}{m} = \frac{J/C}{m} = \frac{J}{C \cdot m} = \frac{N \cdot m}{C \cdot m} = \frac{N}{C}$$

Solved Example Problem

Two parallel charged plates are 2.0×10^{-3} meters apart. If the potential difference between the plates is 50. Volts, what is the electric field intensity between the plates?

$$E = \frac{V}{d}$$

$$E = \frac{50.\ V}{2.0 \times 10^{-3}\ m}$$

$$E = 2.5 \times 10^4\ V/m$$

Unit 10
Practice Questions

1. An electron is located in the electric field between two parallel metal plates as shown in the diagram below.

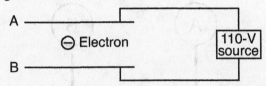

If the electron is attracted to plate A, then plate A is charged

(1) positively, and the electric field is directed from plate A toward plate B
(2) positively, and the electric field is directed from plate B toward plate A
(3) negatively, and the electric field is directed from plate A toward plate B
(4) negatively, and the electric field is directed from plate B toward plate A

---

2. A potential difference of 10.0 volts exists between two points, A and B, within an electric field. What is the magnitude of charge that requires 2.0×10^{-2} joule of work to move it from A to B?

(1) 5.0×10^2 C   (3) 5.0×10^{-2} C
(2) 2.0×10^{-1} C   (4) 2.0×10^{-3} C

3. Which quantity of excess electric charge could be found on an object?

(1) 6.25×10^{-19} C   (3) 6.25 elementary charges
(2) 4.80×10^{-19} C   (4) 1.60 elementary charges

4. The diagram below represents two electrically charged identical-sized metal spheres, A and B.

A   B

$+2.0 \times 10^{-7}$ C $+1.0 \times 10^{-7}$ C

If the spheres are brought into contact, which sphere will have a net gain of electrons?

(1) A, only   (3) both A and B
(2) B, only   (4) neither A nor B

195

5. If the distance separating an electron and a proton is halved, the magnitude of the electrostatic force between these charged particles will be

(1) unchanged   (2) doubled   (3) quartered   (4) quadrupled

6. Two similar metal spheres, A and B, have charges of $+2.0 \times 10^{-6}$ coulomb and $+1.0 \times 10^{-6}$, respectively, as shown in the diagram below.

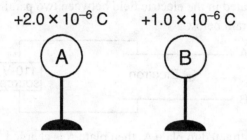

$+2.0 \times 10^{-6}$ C $+1.0 \times 10^{-6}$ C

The magnitude of the electrostatic force on A due to B is 2.4 newtons. What is the magnitude of the electrostatic force on B due to A?

(1) 1.2 N   (2) 2.4 N   (3) 4.8 N   (4) 9.6 N

7. Metal sphere A has a charge of -2 units and an identical metal sphere, B, has a charge of -4 units. If the spheres are brought into contact with each other and then separated, the charge on sphere B will be

(1) 0 units   (2) -2 units   (3) -3 units   (4) +4 units

8. A positively charged glass rod attracts object X. The net charge of object X

(1) may be zero or negative   (3) must be negative
(2) may be zero or positive   (4) must be positive

9. What is the magnitude of the electric field intensity at a point where a proton experiences an electrostatic force of magnitude 2.30×10^{-25} newton?

(1) 3.68×10^{-44} N/C   (3) 3.68×10^{6} N/C
(2) 1.44×10^{-6} N/C   (4) 1.44×10^{44} N/C

10. How much electrical energy is required to move a 4.00-microcoulomb charge through a potential difference of 36.0 volts?

(1) 9.00×10 J   (3) 1.44×10^{-4} J
(2) 144 J   (4) 1.11×10^{-7} J

11. Which graph best represents the relationship between the strength of an electric field and distance from a point charge?

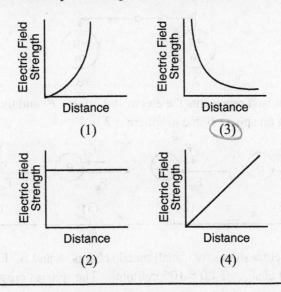

(1)   (3)

(2)   (4)

12. In the diagram below, proton p, neutron n, and electron e are located as shown between two oppositely charged plates.

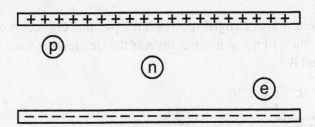

The magnitude of acceleration will be greatest for the

(1) neutron, because it has the greatest mass
(2) neutron, because it is neutral
(3) electron, because it has the smallest mass
(4) proton, because it is farthest from the negative plate

13. Two protons are located one meter apart. Compared to the gravitational force of attraction between the two protons, the electrostatic force between the protons is

(1) stronger and repulsive   (3) stronger and attractive
(2) weaker and repulsive   (4) weaker and attractive

14. The diagram below shows two identical metal spheres, *A* and *B*, separated by distance *d*. Each sphere has mass *m* and possesses charge *q*.

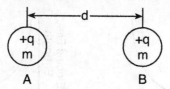

Which diagram best represents the electrostatic force F_e and the gravitational force F_g acting on sphere *B* due to sphere *A* ?

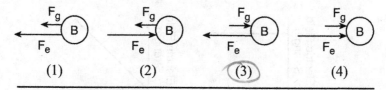

15. The diagram below shows two small metal spheres, A and B. Each sphere possesses a net charge of 4.0×10^{-6} coulomb. The spheres are separated by a distance of 1.0 meter.

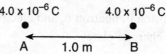

Which combination of charged spheres and separation distance produces an electrostatic force of the same magnitude as the electrostatic force between spheres A and B ?

(1)

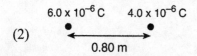

(2)

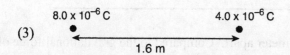

(3)

(4)

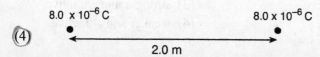

16. The magnitude of the electrostatic force between two point charges is F. If the distance between the charges is doubled, the electrostatic force between the charges will become

   (1) $\frac{F}{4}$ (2) $2F$ (3) $\frac{F}{2}$ (4) $4F$

17. An object with a net charge of 4.80×10^{-6} coulomb experiences an electrostatic force having a magnitude of 6.00×10^{-2} newton when placed near a negatively charged metal sphere. What is the electric field strength at this location?

   (1) 1.25×10^4 N/C directed away from the sphere
   (2) 1.25×10^4 N/C directed toward the sphere
   (3) 2.88×10^{-8} N/C directed away from the sphere
   (4) 2.88×10^{-8} N/C directed toward the sphere

18. The diagram below shows three neutral metal spheres, x, y, and z, in contact and on insulating stands.

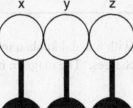

Which diagram best represents the charge distribution on the spheres when a positively charged rod is brought near sphere x, but does not touch it?

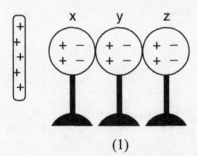

(1)

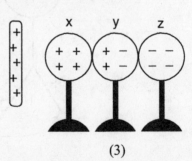

(3)

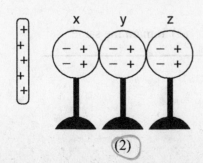

(2)

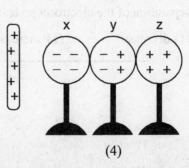

(4)

19. The diagram below shows two identical metal spheres, A and B, on insulated stands. Each sphere possesses a net charge of -3×10^{-6} coulomb.

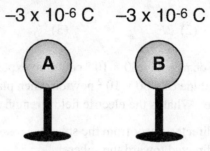

$$-3 \times 10^{-6} \text{ C} \quad -3 \times 10^{-6} \text{ C}$$

If the spheres are brought into contact with each other and then separated, the charge on sphere A will be

(1) 0 C

(2) $+3 \times 10^{-6}$ C

(3) -3×10^{-6} C

(4) -6×10^{-6} C

20. An electroscope is a device with a metal knob, a metal stem, and freely hanging metal leaves used to detect charges. The diagram below shows a positively charged leaf electroscope.

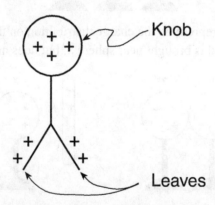

As a positively charged glass rod is brought near the knob of the electroscope, the separation of the electroscope leaves will

(1) decrease   (2) increase   (3) remain the same

UNIT 11
Current Electricity and Circuits

In the previous unit, static electricity, or stationary charges were explored. In this unit, *current electricity*, or the flow of electric charge, will be covered. The conditions necessary for electric current, Ohm's Law, and electric circuits will be discussed.

Electric Current

Electric current refers to moving electric charge. In the previous unit, it was determined that charge is measured in Coulombs (C). The flow of electric current is the charge that passes a certain point per unit time. In equation form:

$$I = \frac{\Delta q}{t}$$

where:
I = current
Δq = amount of charge
t = time

Since charge is measured in Coulombs (C) and time in seconds (s), the units of electric current are Coulombs per second (C/s). Coulombs per second are equivalent to the Ampère (A), or abbreviated the amp. The Ampère is one of the seven base units in the SI system of measurement and is named after André-Marie Ampère.

Electric current can flow in a variety of materials. It is most often thought of as current flowing through a wire in an electric circuit, however, current can also flow through other solids, liquids, and gases.

André-Marie Ampère

André-Marie Ampère was a French physicist and mathematician most noted for discovering Ampère's Law. He proved that two parallel wires with electrical currents will exert forces between them. He called this electrodynamics. He showed that the force was proportional to both the lengths of the wires and the currents in them. The SI unit of current (the Ampère, A) is named after him.

Born: January 20, 1775; France
Died: June 16, 1836; France

Solved Example Problems

1. If 18.0 Coulombs of charge pass a point in an electric circuit in 6.00 seconds, what is the current at that point?

$$I = \frac{\Delta q}{t}$$

$$I = \frac{18.0 \ C}{6.00 \ s}$$

$$I = 3.00 \ A$$

2. If 800. electrons pass a given point in an electrical circuit in 2.00 seconds, the current at this point is

$$q = n \cdot e$$
$$q = (800.) \cdot (1.60 \times 10^{-19} \ C)$$
$$q = 1.28 \times 10^{-16} \ C$$

$$I = \frac{\Delta q}{t}$$
$$I = \frac{1.28 \times 10^{-16} \ C}{2.00 \ s}$$
$$I = 6.40 \times 10^{-17} \ A$$

Current in Solids

In solids, the electron is the particle that will move in an electric current. Historically, however, it was thought that positive charges were flowing to make an electric current. This logic is called *conventional* current, or the flow of positive charge. If the electrons are traveling clockwise around an electric circuit, then the conventional current (think of the *absence of an electron* or *hole*) is traveling counterclockwise. Since metals contain many electrons that are free to move, metals are good conductors of electricity. Materials that do not have many free electrons to flow are not good conductors, and are also known as insulators.

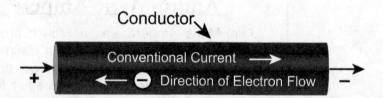

Conductor

Conventional Current →

+   ← ⊖ Direction of Electron Flow   –

Current in Liquids

In liquids, ions (charged atoms or molecules) are the agents that flow to make an electric current. While many any liquids are composed of neutral particles and these liquids are poor conductors of electricity, acids, bases, and dissolved salts have many ions in their solutions and are very good conductors. The dissolved salts and ions in a liquid are called *electrolytes*. Both positive and negative ions can move in a liquid. The negative ions will move in one direction, and the positive ions will move in the opposite direction.

Current in Gases

Like liquids, gases are often comprised of neutral molecules and therefore are poor conductors. Gases may become ionized by exposure to radiation, electric fields, or collisions. An ionized gas, or plasma, will conduct electricity and, like liquids, both the positive and negative ions will flow in opposite directions.

Conditions Necessary for Electric Current

In order for electric current to flow, there are two things that must be present. These two things constitute the most basic of *electric circuits* when connected properly:

1. a source of potential difference (voltage) from a source like a battery or generator, and
2. a path for the electric current to follow, typically from a wire made of a conductor like copper.

Although current will flow if these two things are present, in a real-life setting it would flow only for a very brief amount of time. If one thinks of the battery as a source of *energy* for the electrons that flow in a circuit, there is not really any device in this basic electric circuit to use or convert this energy into any other form of energy. As a result, the current will be very large. The battery will die very quickly, and a large amount of heat will be produced in the wires and the battery itself. The third circuit element that makes the current useful is a **resistor** - a device that will convert the electrical energy into other forms such as mechanical motion, light, sound, heat, or some other form. The wire itself does have some resistance, but it is very small as compared to other devices such as light bulbs or motors.

So, for a practical circuit, the three necessary elements are: potential source, path, and resistor. A common simple circuit consists of a battery, a wire, and a light bulb, as shown below:

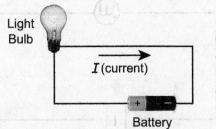

Light Bulb

I (current)

Battery

As seen in the diagram above, the *conventional current*, or movement of positive charge, is *clockwise* around the circuit. The *electron flow*, which exits the negative terminal of the battery, is therefore *counter-clockwise*.

Although the drawing above is very useful in visualizing the circuit, it is time consuming to draw and it relies on a person's drawing skills. As a result, there is a set of symbols that universally represent the different elements of an electric circuit. Some of these symbols that are used to create **schematic diagrams** of circuits are on the following page, as they appear in the Reference Tables:

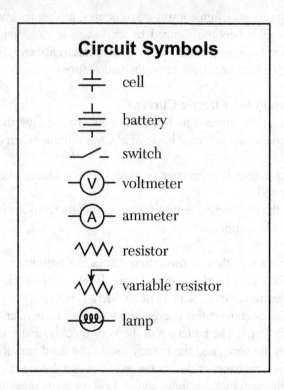

Circuit Symbols

cell

battery

switch

voltmeter

ammeter

resistor

variable resistor

lamp

Using these symbols, the **schematic diagram** of the simple battery and light bulb would look like this:

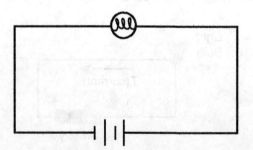

      The schematic above includes the lamp at the top of the diagram, and two cells at the bottom. Multiple cells are called a *battery*. The things we commonly call batteries are technically single cells, the dry cells, such as the size AAA, AA, C, D, or 9V batteries used in personal electronics and many other items. Note that the *longer line of a cell or battery is the positive side*. Therefore, the electron flow in this schematic would be clockwise.

Resistance

In the example circuit on the previous page, a light bulb is the element providing the **resistance** to the circuit. In an ideal circuit the wires provide no resistance. In reality, all elements, including the wires, offer resistance to the circuit. Resistance is the opposition to the flow of electric current in a conductor. There are several factors that affect the flow of electrons as they travel through a conductor:

- **Material:** The effect of the conducting material on the resistance is described by the **resistivity**. The symbol for resistivity is the Greek letter Rho (ρ). The resistance of a conductor is directly proportional to the resistivity. Some resistivities are listed in the table below from the Reference Tables:

Resistivities at 20°C	
Material	Resistivity ($\Omega \cdot$m)
Aluminum	2.82×10^{-8}
Copper	1.72×10^{-8}
Gold	2.44×10^{-8}
Nichrome	$150. \times 10^{-8}$
Silver	1.59×10^{-8}
Tungsten	5.60×10^{-8}

- **Temperature:** There is a direct relationship between the temperature of a metal conductor and resistance. From Chemistry, recall that temperature is the measure of the average kinetic energy of molecules. If a wire were at a relatively high temperature, the metal atoms and molecules would be vibrating more rapidly. As an electron flows from one point to another in a wire, it will interact with the other atoms, molecules, and electrons in the wire more often. These interactions and collisions will result in the electron making many zig-zag movements. These changes in direction and collisions with other particles result in some energy loss of the electron in the circuit. Of course, the energy is not really lost; it is typically converted to heat.

- **Length (L):** The collisions between the electrons and other particles in the wire give rise to increased resistance. With a longer wire, there are more total collisions that will occur, and therefore, more resistance. There is a direct relationship between length of the wire and resistance.

- **Cross-sectional Area (A):** The resistance of a conductor varies inversely with the cross-sectional area. A thicker wire will have a larger cross-sectional area. If one were to take a wire and cut the end of it off, the end would be in the shape of a circle. The formula for area of a circle is A= π r^2.

The formula that relates all of the factors on the previous page (except temperature) from the Reference Tables is:

$$R = \frac{\rho L}{A}$$

The unit of resistance is the Ohm, named for German Physicist Georg Simon Ohm. The symbol for the Ohm is the Greek letter omega (Ω). In terms of radius of the wire, the formula would be:

$$R = \frac{\rho L}{\pi r^2}$$

As you can see from the formula above, by doubling the radius (and thereby doubling the thickness) of a wire, the resistance will become 1/4 as great.

Solved Example Problems

1. Calculate the electrical resistance in a copper wire with a cross-sectional area of 3.0×10^{-5} m^2 that is 15 m long.

$$R = \frac{\rho L}{A}$$

$$R = \frac{(1.72 \times 10^{-8}\,\Omega{:}m)(15\,m)}{3.0 \times 10^{-5}\,m^2}$$

$$R = 8.6 \times 10^{-3}\,\Omega$$

2. Determine the length of an aluminum wire that has an electrical resistance of 4.0×10^{-3} Ω if it has a cross-sectional area of 1.5×10^{-6}m^2.

$$R = \frac{\rho L}{A}$$

$$4.0 \times 10^{-3}\,\Omega = \frac{(2.82 \times 10^{-8}\,\Omega{-}m)L}{1.5 \times 10^{-6}\,m^2}$$

$$L = 0.21\,m$$

Georg Simon Ohm

Georg Simon Ohm was a German physicist, mathematician, and high school teacher. He is most known for discovering Ohm's Law which states that there is a direct relationship between electric potential and current. He conducted his research by working with Volta's electrochemical cell. The SI unit of electrical resistance (the Ohm, Ω) is named after him.

Born: March 16, 1789; Bavaria
Died: July 6, 1854; Bavaria

College Attended: University of Erlangen

Ohm's Law

For many electrical conductors and devices, it was determined that the current through the conductor varies directly with the potential difference. Plotted on a graph with Potential Difference (V) on the y-axis and Current (I) on the x-axis, the graph is linear.

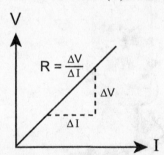

The slope of this graph is the **resistance** of the conductor or electrical device. The equation formed from the slope of this line is the basis of Ohm's Law.

The units of this equation are the units of resistance, the Ohm (Ω). From this equation, it can be noted that one Ohm is equal to 1 Volt per Ampére ($1 \Omega = 1 V/A$). It is important to note that Ohm's Law only applies to *Ohmic devices* (devices that obey Ohm's Law). If the relationship as plotted in the graph is not linear, the resistance of that electrical component is not constant. Some examples of devices that do not obey Ohm's Law include incandescent light bulbs, vacuum tubes, transistors, and gas discharge tubes.

Much like Newton's Second Law, a handy way to work with Ohm's Law is to create an Ohm's Law triangle from the equation. In using the Ohm's Law triangle to the right, cover the variable you are trying to solve for with your finger or thumb. The remaining letters and their arrangement will show whether you need to multiply or divide. For example, solving for current (I) - cover the I and what remains is V/R. This tells you that in order to solve for I, you must divide the potential difference (V) by the resistance (R).

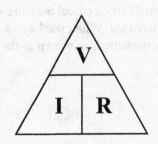

Solved Example Problem
- A 4.00-ohm resistor is connected to a 12.00-volt battery. What current flows in the circuit?

$$R = \frac{V}{I}$$

$$I = \frac{V}{R}$$

$$I = \frac{12.00 \text{ V}}{4.00 \text{ }\Omega}$$

$$I = 3.00 \text{ A}$$

Measurement of Current and Voltage

The device used to measure potential difference (voltage) is called a **voltmeter**. A voltmeter may be *digital* or *analog* and is always connected in parallel to (on either side of) the component to be measured, such as a resistor. Voltmeters have *very high resistance* to prevent the current from traveling through them.

Analog Voltmeter

Digital Voltmeter

The digital voltmeter above is actually a *multimeter*, which can be used to make a variety of electrical measurements, depending on the selection on the dial and how it is connected. When used as a voltmeter, it must be connected in parallel to the device to be measured, as shown in the diagram below:

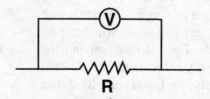

The device used to measure current is called an **ammeter**. As with a voltmeter, this device may also be either digital or analog. An ammeter is always connected in series (in line) with the device or segment of the circuit to be measured. Ammeters have *very low resistance* so they do not add any significant amount of total resistance to the circuit.

Analog Ammeter

Digital Ammeter

Again, ammeters, whether digital or analog, must be connected in series with the device to be measured as illustrated in the diagram below:

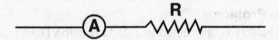

The diagram below shows an ammeter and voltmeter properly connected in a simple circuit:

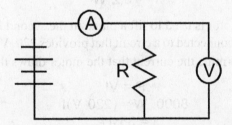

Electrical Power

Recall from the energy unit that **power** is the rate at which work is done. In equation form:

$$P = \frac{W}{t}$$

In the static electricity unit, the electric potential difference was defined as the work per unit charge, or V=W/q. Solved for W:

$$W = qV$$

Substituting this into the power equation yields:

$$P = \frac{qV}{t}$$

Since Current (I) is q/t, this can be simplified:

$$P = VI$$

The equation above is the fundamental equation for electrical power. Like mechanical power, the unit is the Watt (W). The equation for power can be restated in terms of Resistance as well if Ohm's Law (V=IR or I=V/R) is substituted into the equation above. As a result, there are three ways to express electrical power:

$$P = VI = I^2R = \frac{V^2}{R}$$

Solved Example Problems

1. A 9.0 V battery is connected to a 25 Ω resistor. Determine the amount of power developed by the resistor.

$$P = \frac{V^2}{R}$$
$$P = \frac{(9.0 \ V)^2}{25 \ \Omega}$$
$$P = 3.2 \ W$$

2. A 8000. W motor is used to lift a piano to the second floor of a building. The motor is connected to a circuit that provides 220. V of electric potential.

a. Determine the current that the motor draws through this circuit.

$$P = VI$$
$$8000. \ W = (220. \ V)I$$
$$I = 36.4 \ A$$

b. Calculate the electrical resistance of the motor.

$$P = I^2R$$
$$8000. \ W = (36.4 \ A)^2 R$$
$$R = 6.04 \ \Omega$$

More than one equation could have been used to solve for the resistance in (b) above, such as R=V/I or P=V²/R.

Electrical Energy

Electrical energy is the work that is done in an electric circuit. It can also represent the heat generated by an electrical device. Much like the energy unit, multiplying the power by time yields the energy. In the electricity unit, electrical energy has the same symbol in a formula as work, W:

$$W = Pt$$

Subsequently, each of the formulas for power can be multiplied by time to give equations for energy:

$$W = Pt = VIt = I^2Rt = \frac{V^2t}{R}$$

The unit of electrical energy is the Joule (J). One Joule is equivalent to one Watt·second. Another unit of measure for electrical energy is the *kilowatt·hour* (kWh).

$$1\,kWh = 3.6 \times 10^6\,J$$

This is often the unit of measure that utility companies use to bill their customers for energy (electricity) usage. A common misnomer is to call such a company "The Power Company" because, in fact, it is selling you energy, not power!

Solved Example Problem

- A toaster connected to a 120. V circuit draws a current of 5.5 A. How much electrical energy does the toaster use in 60.0 s?

$$W = VIt$$
$$W = (120.\,V)(5.5\,A)(60.0\,s)$$
$$W = 39600\,J$$

Series Circuits:(only one path for current)

A series circuit is a circuit containing multiple circuit devices connected so that the current passes through each and every device without branching off. If the resistors in a series circuit were light bulbs, for example, and one of the bulbs were to blow out, the current in the entire circuit would stop and all bulbs would go out.

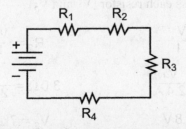

There are only a few real-world situations where series circuits are in use. Certain strings of holiday lights, for example, are wired in series. It is cheaper to wire a string of lights in series because it uses less wire to construct. A drawback, however, is that when one bulb burns out, the entire string goes out, and it is difficult to locate the burned out bulb. Some properties of series circuits are listed below.

1. The current is the same in every part of the circuit and through each resistor: ($I = I_1 = I_2 = I_3 = ...$)
2. Total (Equivalent) resistance is equal to the sum of the individual resistances: ($R_{eq} = R_1 + R_2 + R_3 + ...$)
3. Total voltage is equal to the sum of the voltages across the resistances: ($V = V_1 + V_2 + V_3 + ...$)
4. As you add additional resistors to the circuit
   a. The total (equivalent) resistance increases
   b. The total voltage is constant but the voltage across each resistance decreases
   c. The total current (there is only one) decreases
   d. The power expended decreases

Solved Example Problem

• Two resistors, 2.0 Ω and 3.0 Ω, are connected in series with a 12 V battery.

Calculate:

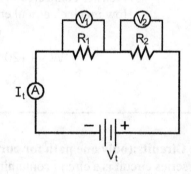

a. the equivalent resistance of the circuit (R_{eq})

$$R_{eq} = R_1 + R_2$$
$$R_{eq} = 2.0\,\Omega + 3.0\,\Omega$$
$$R_{eq} = 5.0\,\Omega$$

b. the current leaving the battery (I)

$$R_{eq} = \frac{V_t}{I}$$

$$5.0\,\Omega = \frac{12\,V}{I}$$

$$I = 2.4\,A$$

c. the current through each resistor (I_1 and I_2)

$$I = I_1 = I_2 = 2.4\,A$$

d. the voltage drop across each resistor (V_1 and V_2)

$$R_1 = \frac{V_1}{I} \qquad\qquad R_2 = \frac{V_2}{I}$$

$$2.0\,\Omega = \frac{V_1}{2.4\,A} \qquad\qquad 3.0\,\Omega = \frac{V_2}{2.4\,A}$$

$$V_1 = 4.8\,V \qquad\qquad V_2 = 7.2\,V$$

Parallel Circuits:(more than one path for the current)

Unlike series circuits that have only one path for the current to travel, a parallel circuit has multiple paths, or branches. Therefore, the current is split among the branches. Each electron in the current is only passing through one branch before returning to the source of potential. Therefore, the potential drop (voltage) across each branch is equal to the total from the source. The total current is measured from the source of potential and each branch's current combined is equal to the total.

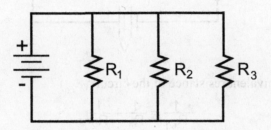

If one bulb in a multi-bulb parallel circuit burns out, the other bulbs will remain lit, but they do not get brighter. The total current will decrease, but the current through each branch remains the same. Some strings of holiday lights are parallel circuits. It is more costly to manufacture as it requires more wire to construct, but it is very easy to pinpoint a burned out bulb, as it will be the only one not lit. Some properties of parallel circuits are listed below.

1. The voltage across each branch is the same: $(V = V_1 = V_2 = V_3 = ...)$

2. The total current is equal to the sum of the currents through each of the branches: $(I = I_1 + I_2 + I_3 + ...)$

3. As resistances are added the total (equivalent) resistance decreases, therefore the total current and power expended increases:

$$\frac{1}{R_{eq}} = \frac{1}{R_1} + \frac{1}{R_2} + \frac{1}{R_3} + ...$$

Note that the equivalent resistance is always smaller than the smallest resistor in the circuit.

4. Adding resistances does not affect the individual currents and voltages in each branch.

Solved Example Problem

• Two resistors, 2.0 Ω and 3.0 Ω, are connected in parallel with a 12 V battery.

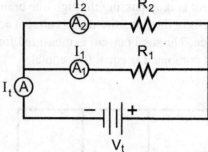

Calculate:

a. the equivalent resistance of the circuit (R_{eq})

$$\frac{1}{R_{eq}} = \frac{1}{R_1} + \frac{1}{R_2}$$

$$\frac{1}{R_{eq}} = \frac{1}{2.0\,\Omega} + \frac{1}{3.0\,\Omega}$$

$$R_{eq} = 1.2\,\Omega$$

b. the current leaving the battery (I_t)

$$R_{eq} = \frac{V_t}{I_t}$$

$$1.2\,\Omega = \frac{12\,V}{I_t}$$

$$I_t = 10.\,A$$

c. the current through each resistor (I_1 and I_2)

$$R_1 = \frac{V}{I_1} \qquad R_2 = \frac{V}{I_2}$$

$$2.0\,\Omega = \frac{12\,V}{I_1} \qquad 3.0\,\Omega = \frac{12\,V}{I_2}$$

$$I_1 = 6.0\,A \qquad I_2 = 4.0\,A$$

d. the voltage drop across each resistor (V_1 and V_2)

$$V = V_1 = V_2 = 12\,V$$

Combination Series-Parallel Circuits

Another type of basic circuit is a combination of series and parallel circuit elements. An example of this type of circuit would be some of the more modern strings of holiday lights. With a wiring configuration like this, if one bulb burns out, any other bulbs in series with that bulb would also go out, but the rest of the string would remain lit. Therefore only a section would go out and not the entire string. This gives some of the advantages of a parallel circuit while using less wire. A diagram of the simplest type of series-parallel combination circuit is below.

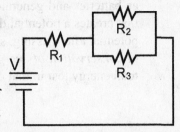

This type of circuit is somewhat advanced, and the best way to analyze it is to first find the equivalent resistance of the entire circuit. In order to do this, parts of the circuit must be simplified as series or parallel. Starting with resistors R_2 and R_3, one can see that these two resistors are in parallel. The equivalent resistance of just these two resistors can be determined. Once that it found, this equivalent resistance is in series with resistor R_1, and the equivalent resistance for the entire circuit can be found. From there, other aspects of the circuit can be analyzed. It is important to note that the voltage across R_2 and R_3 will be the same as each other, but not the same as the battery.

Kirchhoff's Rules

Named after German Physicist Gustav Kirchhoff who explained two different ways of analyzing an electric circuit in 1845. These two methods utilize the concepts of conservation of energy, and conservation of charge.

- **Junction Rule:** Kirchhoff's junction rule states that the total amount of electric current entering a junction must equal the total electric current leaving a junction. A junction is defined as a point in a circuit where two or more wires come together. This is a statement of *conservation of charge* - charge (and therefore current) can be neither created nor destroyed. See the example below:

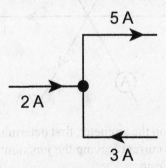

In the diagram above, note that the total current entering the junction (the point denoted with a "dot" in the circuit segment) is 5A (2A + 3A), and the total current leaving the junction is also 5A. Therefore the total charge is conserved at this junction.

- **Loop Rule:** Kirchhoff's loop rule states that the total of the potential rises in a circuit must equal the total of the potential drops. A device in a circuit that creates a potential rise is a device that gives the charges a higher potential, such as batteries and generators (when connected in the typical fashion). A device that creates a potential drop is a device that causes the charges to have a lower potential after passing, such as resistors. Kirchhoff's loop rule is an application of *conservation of energy* - the total energy gained in a circuit must equal the total energy lost in that circuit.

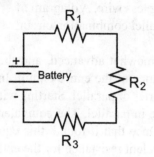

Kirchhoff's loop rule, applied to the circuit above, states that the total potential rise must equal the total potential drop. In this circuit, the total potential rise is the potential (voltage) of the battery. The total potential drop is the sum of the voltages across each of the resistors above.

Solved Example Problem

- For the circuit segment below, determine the reading on ammeter A.

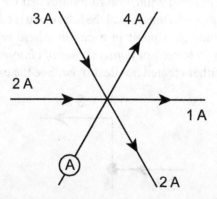

To determine the reading on the ammeter, first determine the total current entering the junction, and the total current leaving the junction:

Entering:	Leaving:
2A+3A=5A	4A+1A+2A=7A

For the total current entering and leaving the junction to be the same, there must be an additional **2A of current going into the junction**.

Unit 11
Practice Questions

1. At 20°C, four conducting wires made of different materials have the same length and the same diameter. Which wire has the least resistance?

   (1) aluminum   (2) gold   (3) nichrome   (4) tungsten

2. Three identical lamps are connected in parallel with each other. If the resistance of each lamp is X ohms, what is the equivalent resistance of this parallel combination?

   (1) $X\ \Omega$ (2) $\dfrac{X}{3}\ \Omega$ (3) $3X\ \Omega$ (4) $\dfrac{3}{X}\ \Omega$

3. In the electric circuit diagram below, possible locations of an ammeter and a voltmeter are indicated by circles 1, 2, 3, and 4.

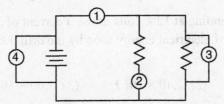

   Where should an ammeter be located to correctly measure the total current and where should a voltmeter be located to correctly measure the total voltage?

   (1) ammeter at 1 and voltmeter at 4   (3) ammeter at 3 and voltmeter at 4
   (2) ammeter at 2 and voltmeter at 3   (4) ammeter at 1 and voltmeter at 2

4. What is the current in a 100.-ohm resistor connected to a 0.40-volt source of potential difference?

   (1) 250 mA   (2) 40. mA   (3) 2.5 mA   (4) 4.0 mA

5. A 150-watt lightbulb is brighter than a 60.-watt lightbulb when both are operating at a potential difference of 110 volts. Compared to the resistance of and the current drawn by the 150-watt lightbulb, the 60.-watt lightbulb has

   (1) less resistance and draws more current
   (2) less resistance and draws less current
   (3) more resistance and draws more current
   (4) more resistance and draws less current

6. What is the minimum equipment needed to determine the power dissipated in a resistor of unknown value?

    (1) a voltmeter, only
    (2) an ammeter, only
    (3) a voltmeter and an ammeter, only
    (4) a voltmeter, an ammeter, and a stopwatch

7. A circuit consists of a resistor and a battery. Increasing the voltage of the battery while keeping the temperature of the circuit constant would result in an increase in

    (1) current, only
    (2) resistance, only
    (3) both current and resistance
    (4) neither current nor resistance

8. Charge flowing at the rate of 2.50×10^{16} elementary charges per second is equivalent to a current of

    (1) 2.50×10^{13} A
    (3) 4.00×10^{-3} A
    (2) 6.25×10^{5} A
    (4) 2.50×10^{-3} A

9. An electric drill operating at 120. volts draws a current of 3.00 amperes. What is the total amount of electrical energy used by the drill during 1.00 minute of operation?

    (1) 2.16×10^{4} J
    (2) 2.40×10^{3} J
    (3) 3.60×10^{2} J
    (4) 4.00×10^{1} J

10. A 6.0-ohm lamp requires 0.25 ampere of current to operate. In which circuit below would the lamp operate correctly when switch S is closed?

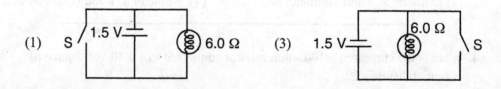

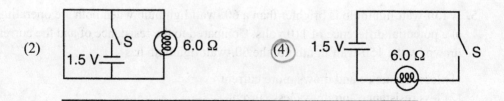

CURRENT ELECTRICITY AND CIRCUITS

11. What is the total current in a circuit consisting of six operating 100-watt lamps connected in parallel to a 120-volt source?

(1) 5 A

(3) 600 A

(2) 20 A

(4) 12 000 A

12. A 4.50-volt personal stereo uses 1950 joules of electrical energy in one hour. What is the electrical resistance of the personal stereo?

(1) 433 Ω

(2) 96.3 Ω

(3) 37.4 Ω

(4) 0.623 Ω

13. Which quantity and unit are correctly paired?

(1) resistivity and $\frac{\Omega}{m}$

(3) current and C•s

(2) potential difference and eV

(4) electric field strength and $\frac{N}{C}$

14. What is the resistance at 20.°C of a 2.0-meter length of tungsten wire with a cross-sectional area of 7.9×10^{-7} meter^2?

(1) 5.7×10^{-1} Ω

(2) 1.4×10^{-1} Ω

(3) 7.1×10^{-2} Ω

(4) 4.0×10^{2} Ω

15. Two identical resistors connected in series have an equivalent resistance of 4 ohms. The same two resistors, when connected in parallel, have an equivalent resistance of

(1) 1 Ω

(2) 2 Ω

(3) 8 Ω

(4) 4 Ω

16. As the number of resistors in a parallel circuit is increased, what happens to the equivalent resistance of the circuit and total current in the circuit?

(1) Both equivalent resistance and total current decrease.

(2) Both equivalent resistance and total current increase.

(3) Equivalent resistance decreases and total current increases.

(4) Equivalent resistance increases and total current decreases.

17. The graph below represents the relationship between the potential difference (V) across a resistor and the current (I) through the resistor.

Through which entire interval does the resistor obey Ohm's law?

(1) AB

(2) BC

(3) CD

(4) AD

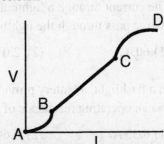

18. What must be inserted between points A and B to establish a steady electric current in the incomplete circuit represented in the diagram below?

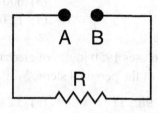

(1) switch
(2) voltmeter

(3) magnetic field source
(4) source of potential difference

_____

19. An electrical generator in a science classroom makes a lightbulb glow when a student turns a hand crank on the generator. During its operation, this generator converts

(1) chemical energy to electrical energy
(2) mechanical energy to electrical energy
(3) electrical energy to mechanical energy
(4) electrical energy to chemical energy

20. Which changes would cause the greatest increase in the rate of flow of charge through a conducting wire?

(1) increasing the applied potential difference and decreasing the length of wire
(2) increasing the applied potential difference and increasing the length of wire
(3) decreasing the applied potential difference and decreasing the length of wire
(4) decreasing the applied potential difference and increasing the length of wire

21. The resistance of a 60.-watt lightbulb operated at 120 volts is approximately

(1) 720 Ω   (2) 240 Ω   (3) 120 Ω   (4) 60. Ω

22. The current through a lightbulb is 2.0 amperes. How many coulombs of electric charge pass through the lightbulb in one minute?

(1) 60. C   (2) 2.0 C   (3) 120 C   (4) 240 C

23. In a flashlight, a battery provides a total of 3.0 volts to a bulb. If the flashlight bulb has an operating resistance of 5.0 ohms, the current through the bulb is

(1) 0.30 A   (2) 0.60 A   (3) 1.5 A   (4) 1.7 A

24. A complete circuit is left on for several minutes, causing the connecting copper wire to become hot. As the temperature of the wire increases, the electrical resistance of the wire

    (1) decreases   (2) increases   (3) remains the same

25. A 100.-ohm resistor and an unknown resistor are connected in series to a 10.0-volt battery. If the potential drop across the 100.-ohm resistor is 4.00 volts, the resistance of the unknown resistor is

    (1) 50.0 Ω   (2) 100. Ω   (3) 150. Ω   (4) 200. Ω

26. In the circuit diagram shown below, ammeter A_1 reads 10. amperes.

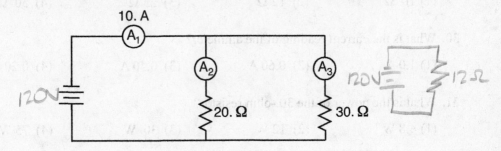

    What is the reading of ammeter A_2?

    (1) 6.0 A   (2) 10. A   (3) 20. A   (4) 4.0 A

27. The current traveling from the cathode to the screen in a television picture tube is 5.0×10^{-5} ampere. How many electrons strike the screen in 5.0 seconds?

    (1) 3.1×10^{24} (2) 6.3×10^{18} (3) 1.6×10^{15} (4) 1.0×10^{5}

28. The table below lists various characteristics of two metallic wires, A and B.

Wire	Material	Temperature (°C)	Length (m)	Cross-Sectional Area (m²)	Resistance (Ω)
A	silver	20.	0.10	0.010	R
B	silver	20.	0.20	0.020	???

    If wire A has resistance R, then wire B has resistance

    (1) R   (2) 2R   (3) $\dfrac{R}{2}$ (4) 4R

Base your answers to questions 29-31 on the information and diagram below.

A 20.-ohm resistor and a 30.-ohm resistor are connected in parallel to a 12-volt battery as shown. An ammeter is connected as shown.

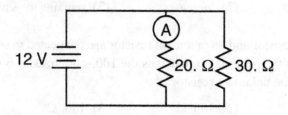

29. What is the equivalent resistance of the circuit?

(1) 10. Ω   (2) 12 Ω   (3) 25 Ω   (4) 50. Ω

30. What is the current reading of the ammeter?

(1) 1.0 A   (2) 0.60 A   (3) 0.40 A   (4) 0.20 A

31. What is the power of the 30.-ohm resistor?

(1) 4.8 W   (2) 12 W   (3) 30. W   (4) 75 W

---

32. In which circuit would ammeter A show the greatest current?

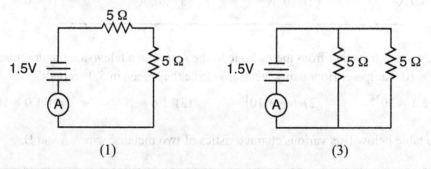

(1)   (3)

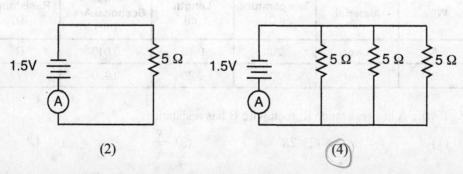

(2)   (4)

Base your answers to questions 33-34 on the circuit diagram below.

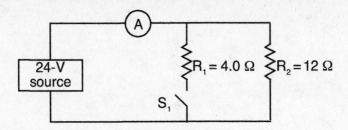

33. If switch S_1 is open, the reading of ammeter A is

    (1) 0.50 A   (2) 2.0 A   (3) 1.5 A   (4) 6.0 A

34. If switch S_1 is closed, the equivalent resistance of the circuit is

    (1) 8.0 Ω   (2) 2.0 Ω   (3) 3.0 Ω   (4) 16 Ω

---

35. In the diagram below, lamps L_1 and L_2 are connected to a constant voltage power supply.

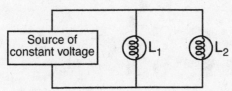

If lamp L_1 burns out, the brightness of L_2 will

    (1) decrease   (2) increase   (3) remain the same

NOTES:

UNIT 12
Electromagnetism

Electricity and magnetism are interrelated. Magnetic forces and magnetic fields are created by moving charges, and charges can be made to move by magnets and magnetic fields. This is why the topic is called *Electromagnetism* and not magnetism.

Permanent Magnets

The most common type of magnets found in everyday life is the permanent magnet. Permanent magnets include bar magnets, horseshoe magnets, and refrigerator magnets. A **permanent magnet** is made from a material that can retain a magnetic field for very long periods of time. Materials that can be magnetized and can make good permanent magnets are called **ferromagnetic materials**. Some examples of ferromagnetic materials include iron, cobalt, and nickel. Most permanent magnets are made of alloys, such as steel and alnico, that include ferromagnetic materials. Softer materials, such as soft iron, make temporary magnets. Magnetic properties of temporary magnets are exhibited only in the presence of other magnets.

Magnetic Domains
A **magnetic domain** is a region in a material in which the magnetic properties are uniform. If most or all of the domains in a ferromagnetic material are randomly oriented, the material is a weak magnet, or non-magnetic as shown in the left diagram below. If most or all of the domains in a ferromagnetic material are oriented in the same direction, the entire object exhibits magnetic properties and can be considered a strong magnet as shown in the right diagram below.

Domains Before
Magnetization

Domains After
Magnetization

The Compass

A compass is a small, permanent magnet that can rotate freely. A compass will align itself in a magnetic field. The colored end of a compass, usually red or blue, is the North Pole of this small magnet and the other end (usually uncolored) is the South Pole. This North Pole of the compass will point in the same direction of a magnetic field.

Some Conventions About Magnets and Magnetic Poles
- The end of a magnet or compass that points toward the North Pole of Earth is called the North Pole of the magnet (opposite is called the South Pole).
- All magnets have both a North and a South Pole – break a magnet in half and you have two smaller magnets!
- Like poles repel and unlike poles attract.

Earth as a Magnet
It is well known that Earth has poles - North and South. The North Pole (in northern Canada) will attract the North Pole of a magnet. So, that means that Earth's north pole is actually like the south pole of a giant bar magnet. The North and South poles of Earth are actually shifting with time due to the movements of the molten layers of Earth. Also, the magnetic North Pole (where a compass would lead you) is actually about 1800 km away from the geographic North Pole of Earth (the point of the axis of rotation of Earth).

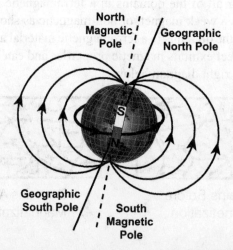

Magnetic Fields

In previous units, the concepts of gravitational and electrical fields were discussed. Much like these type of fields, magnetic fields exist in the vicinity of magnets. A magnetic field (sometimes called a *B-field*) represents the region of force that is exerted on a compass or ferromagnetic material. A magnetic field can be shown by using a small compass or iron filings.

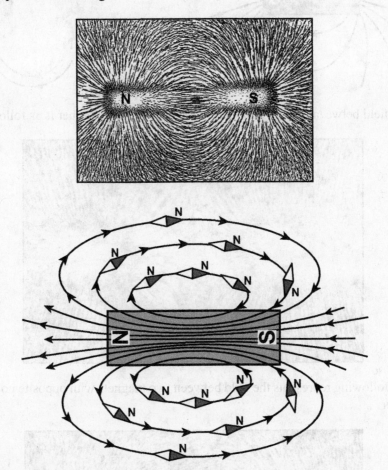

The following are the rules for magnetic fields:
- Field lines never cross.
- The relative number and spacing of the field lines represents relative strength of the field. Magnetic Flux is measured with the Weber (Wb).
- Magnetic Flux Density (B) is the number of flux lines per unit area passing through a plane perpendicular to the direction of the lines. This is measured in Wb/m^2, or Tesla (T).
- Tesla (T) : $1T = 1Wb/m^2$.
- Field lines are a closed curve – lines go in the direction of N to S Pole outside the magnet and they complete their path inside the magnet form S to N Pole.

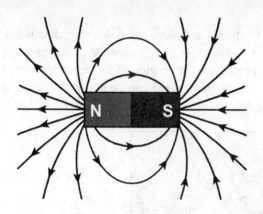

The field between two magnets with like poles near each other is as follows:

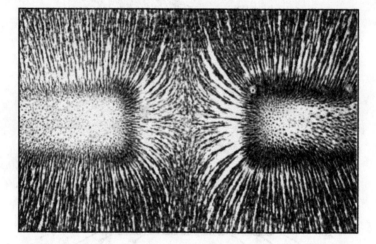

The following represents the field between two magnets with opposite poles facing each other:

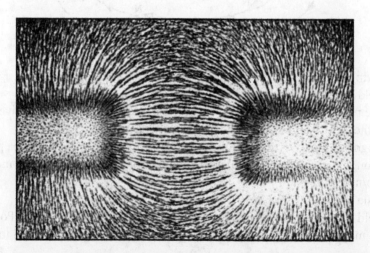

Electromagnetism

In the early 1800's Hans Christian Oersted realized that an electric current moving through a wire deflected a compass. This led to the conclusion that moving electric charges can produce a magnetic field. This can easily be proven by placing a compass or iron filings in a plane perpendicular to a current carrying wire.

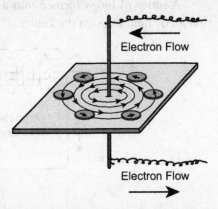

Electron Flow

Electron Flow

The magnetic lines of force (magnetic field lines) that surround a straight current-carrying conductor are concentric circles around that conductor in a plane that is perpendicular to the conductor. The field is strongest near the wire and weakens as the distance from the wire increases. If the direction of the electron flow is known, a *left hand rule* can be used to determine the direction of the magnetic field lines. If the conventional current is given (the movement of positive charge), the right hand can be used instead of the left.

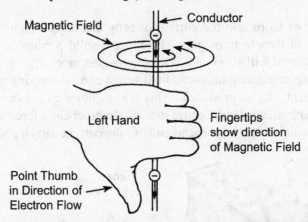

Magnetic Field — Conductor

Left Hand — Fingertips show direction of Magnetic Field

Point Thumb in Direction of Electron Flow

If the wire is bent into a loop, the effects from all parts of the loop will result in a concentrated magnetic field directed through the center of the loop. The field is strongest inside the loop and weaker outside the loop.

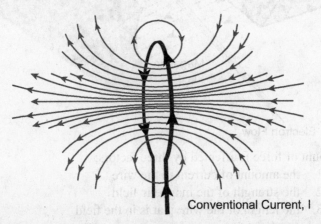

Conventional Current, I

A series of loops formed into a coil, called a *solenoid*, will produce a very strong magnetic field through the center of the coil. This is the basis of an ***electromagnet.***

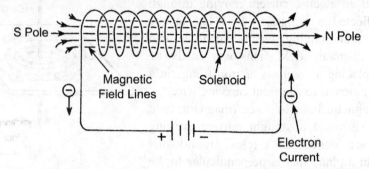

There are three factors that will affect the strength of an electromagnet:
1. the number of loops
2. the amount of current
3. the presence of a permeable core

The number of loops and the current directly affect the magnetic field strength and the strength of the electromagnet. Inserting a solid permeable core made of a ferromagnetic material will also make an electromagnet much stronger.

By combining an external magnetic field with a current carrying wire, a force, and therefore movement, can be produced. This is sometimes called the ***motor principle*** as it explains the physics behind electric motors. In order for a force to be created, the external magnetic field must be perpendicular to the current carrying wire.

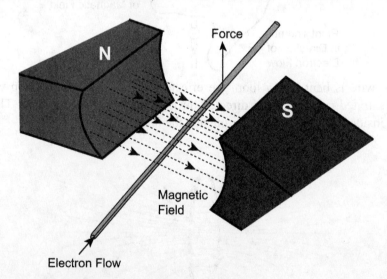

The amount of force is affected by three factors:
1. the amount of current in the wire
2. the strength of the magnetic field
3. the length of the wire that is in the field

Each of these factors directly affects the amount of force produced from this effect. Increasing any or all of them will result in a larger force and, if it is an electric motor, greater acceleration and faster rotation.

A force will also be produced on free-moving charges perpendicular to a magnetic field. Rather than the force being affected by the amount of current and length of the wire, it is now affected by the velocity of the charge and the amount of charge, along with the strength of the field.

If moving electric charges and a magnetic field can produce a force and movement, the opposite should be true as well. If a magnetic field and a conducting wire are moving relative to each other, then an electric current must be produced. This is sometimes called the *generator principle*, or *electromagnetic induction*. In 1831 Michael Faraday discovered that a potential difference (called electromotive force, or emf) is induced when a conductor moves in such a way that it is perpendicular to the magnetic field. If the conductor is part of a complete circuit, then a current is induced in the circuit.

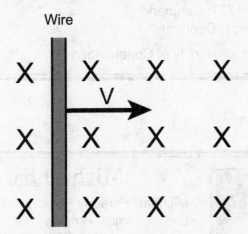

Magnetic field directed into page

In the diagram above, a segment of a wire is moving to the right with a velocity, v. There is an external magnetic field that is directed perpendicularly into the page of this book. As a result, there will be an induced voltage, or emf. The amount of voltage, and current if in a complete circuit, depends on three factors:

1. the speed of the wire movement
2. the strength of the magnetic field
3. the length of the wire that cuts across the field

Each of these factors directly affects the amount of voltage induced. Increasing any or all of these factors will result in a larger emf, and larger current generated if it is part of a complete circuit.

Hans Christian Ørsted

Hans Christian Ørsted was a Dutch chemist and physicist known for his research in electromagnetism. In his lab he observed that a compass needle was deflected from pointing toward north when it was placed near an electric current. As a chemist, he was the first person to produce pure aluminum.

Born: August 14, 1777; Denmark
Died: March 9, 1851; Denmark

College Attended: University of Copenhagen

Michael Faraday

Michael Faraday was an English physicist and chemist known for his work with electromagnetism. His research established the basis for the concept of the electromagnetic field. He coined the terms anode, cathode, electrode, and ion, and successfully discovered benzene. Faraday's ice pail experiment established the concept of the Faraday Cage - the fact that charge resides on the outside of a conductor and the principle behind one's safety in a vehicle if struck by lightning.

Born: September 22, 1791; England
Died: August 25, 1867; England

Unit 12
Practice Questions

1. The diagram below represents a 0.5-kilogram bar magnet and a 0.7-kilogram bar magnet with a distance of 0.2 meter between their centers.

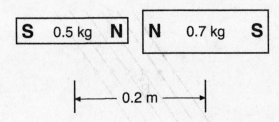

Which statement best describes the forces between the bar magnets?
(1) Gravitational force and magnetic force are both repulsive.
(2) Gravitational force is repulsive and magnetic force is attractive.
(3) Gravitational force is attractive and magnetic force is repulsive.
(4) Gravitational force and magnetic force are both attractive.

2. The diagram below shows the lines of magnetic force between two north magnetic poles.

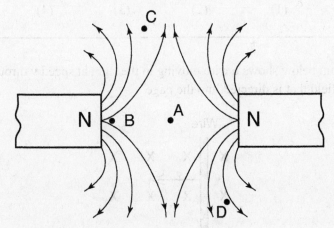

At which point is the magnetic field strength greatest?

(1) A (2) B (3) C (4) D

3. Which type of field is present near a moving electric charge?

(1) an electric field, only
(2) a magnetic field, only
(3) both an electric field and a magnetic field
(4) neither an electric field nor a magnetic field

4. The diagram below represents the magnetic field near point P.

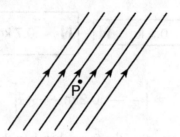

If a compass is placed at point P in the same plane as the magnetic field, which arrow represents the direction the north end of the compass needle will point?

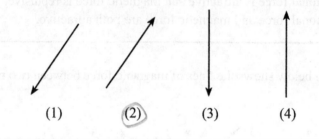

(1)   (2)   (3)   (4)

5. The diagram below shows a wire moving to the right at speed v through a uniform magnetic field that is directed into the page.

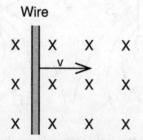

Magnetic field directed into page

As the speed of the wire is increased, the induced potential difference will

(1) decrease   (2) increase   (3) remain the same

6. The diagram below shows a bar magnet.

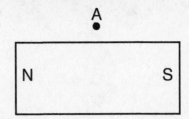

Which arrow best represents the direction of the needle of a compass placed at point A?

(1) ↑   (2) ↓   (3) →   (4) ←

7. Which diagram best represents magnetic flux lines around a bar magnet?

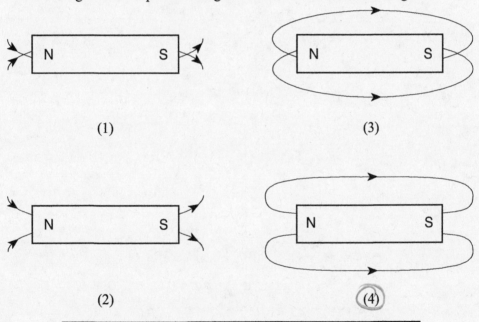

(1)

(3)

(2)

(4)

NOTES:

UNITS 10-12 Cumulative Review
Part A and B1

1. The diagram below shows a beam of electrons fired through the region between two oppositely charged parallel plates in a cathode ray tube.

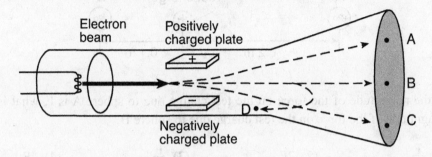

After passing between the charged plates, the electrons will most likely travel path
(1) A   (2) B   (3) C   (4) D

2. If 1.0 joule of work is required to move 1.0 coulomb of charge between two points in an electric field, the potential difference between the two points is

(1) 1.0×10^0 V   (3) 6.3×10^{18} V

(2) 9.0×10^9 V   (4) 1.6×10^{-19} V

3. In the diagram below, P is a point near a negatively charged sphere.

Which vector best represents the direction of the electric field at point P?

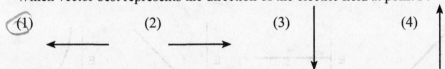

(1)   (2)   (3)   (4)

4. What is the net electrical charge on a magnesium ion that is formed when a neutral magnesium atom loses two electrons?

(1) -3.2×10^{-19} C   (3) $+1.6 \times 10^{-19}$ C

(2) -1.6×10^{-19} C   (4) $+3.2 \times 10^{-19}$ C

5. Oil droplets may gain electrical charges as they are projected through a nozzle. Which quantity of charge is not possible on an oil droplet?

(1) 8.0×10^{-19} C   (3) 3.2×10^{-19} C

(2) 4.8×10^{-19} C   (4) 2.6×10^{-19} C

6. In the diagram below, a positive test charge is located between two charged spheres, A and B. Sphere A has a charge of +2q and is located 0.2 meter from the test charge. Sphere B ha a charge of –2q and is located 0.1 meter from the test charge.

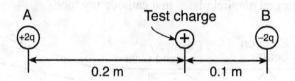

If the magnitude of the force on the test charge due to sphere A is F, what is the magnitude of the force on the test charge due to sphere B?

(1) $\frac{F}{4}$ (2) 2F   (3) $\frac{F}{2}$ (4) 4F

7. A negatively charged plastic comb is brought close to, but does not touch, a small piece of paper. If the comb and the paper are attracted to each other, the charge on the paper

(1) may be negative or neutral   (3) must be negative
(2) may be positive or neutral   (4) must be positive

8. Which graph best represents the relationship between the magnitude of the electric field strength, E, around a point charge and the distance, r, from the point charge?

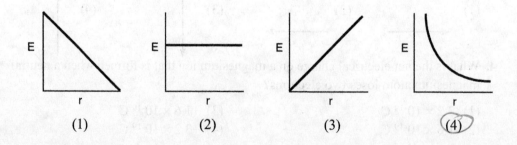

9. In the diagram below, two positively charged spheres, A and B, of masses m_A and m_B are located a distance d apart.

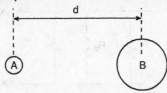

Which diagram best represents the direction of the gravitational force, F_g, and the electrostatic force F_e, acting on sphere A due to the mass and charge of sphere B? (Vectors are not drawn to scale.)

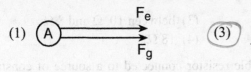

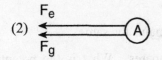

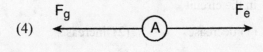

10. When a neutral metal sphere is charged by contact with a positively charged glass rod, the sphere

(1) loses electrons   (2) gains electrons   (3) loses protons   (4) gains protons

11. If the charge on each of two small charged metal spheres is doubled and the distance between the spheres remains fixed, the magnitude of the electric force between the spheres will be

(1) the same   (3) one-half as great
(2) two times as great   (4) four times as great

12. A 2.0-ohm resistor and a 4.0-ohm resistor are connected in series with a 12-volt battery. If the current through the 2.0-ohm resistor is 2.0 amperes, the current through the 4.0-ohm resistor is

(1) 1.0 A   (2) 2.0 A   (3) 3.0 A   (4) 4.0 A

13. In which circuit would current flow through resistor R_1, but *not* through resistor R_2 while switch S is open?

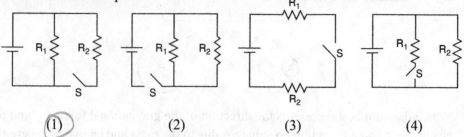

    (1)   (2)   (3)   (4)

14. Three resistors, 4 ohms, 6 ohms, and 8 ohms, are connected in parallel in an electric circuit. The equivalent resistance of the circuit is

    (1) less than 4 Ω   (3) between 10. Ω and 8 Ω
    (2) between 4 Ω and 8 Ω   (4) 18 Ω

15. An electric circuit contains a variable resistor connected to a source of constant voltage. As the resistance of the variable resistor is increased, the power dissipated in the circuit

    (1) decreases   (2) increases   (3) remains the same

16. The current through a 10.-ohm resistor is 1.2 amperes. What is the potential difference across the resistor?

    (1) 8.3 V   (2) 12 V   (3) 14 V   (4) 120 V

17. A copper wire of length L and cross-sectional area A has a resistance R. A second copper wire at the same temperature has a length of 2L and a cross-sectional area of $\frac{1}{2}$A. What is the resistance of the second copper wire?

    (1) R   (2) 2R   (3) $\frac{1}{2}$ (4) 4R

18. The diagram below represents a simple circuit consisting of a variable resistor, a battery, an ammeter, and a voltmeter.

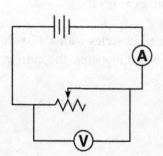

What is the effect of increasing the resistance of the variable resistor from 1000 Ω to 10000 Ω? [Assume constant temperature.]

    (1) The ammeter reading decreases.
    (2) The ammeter reading inreases.
    (3) The voltmeter reading decreases.
    (4) The voltmeter reading increases.

19. In which circuit represented below are meters properly connected to measure the current through resistor R_1 and the potential difference across resistor R_2?

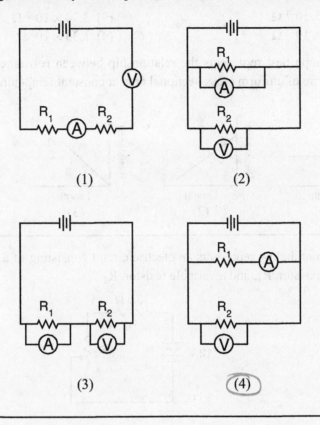

20. A 50-watt light bulb and a 100-watt light bulb are each operated at 110 volts. Compared to the resistance of the 50-watt bulb, the resistance of the 100-watt bulb is

(1) half as great   (3) one-fourth as great

(2) twice as great   (4) four times as great

21. A device operating at a potential difference of 1.5 volts draws a current of 0.20 ampere. How much energy is used by the device in 60. seconds?

(1) 4.5 J   (2) 8.0 J   (3) 12 J   (4) 18 J

22. Pieces of aluminum, copper, gold, and silver wire each have the same length and the same cross-sectional area. Which wire has the lowest resistance at 20°C?

(1) aluminum   (2) copper   (3) gold   (4) silver

23. What is the resistance at 20°C of a 1.50-meter-long aluminum conductor that has a cross-sectional area of 1.13×10^{-6} meter²?

(1) 1.87×10^{-3} Ω

(2) 2.28×10^{-2} Ω

(3) 3.74×10^{-2} Ω

(4) 1.33×10^{6} Ω

24. Which graph best represents the relationship between resistance and length of a copper wire of uniform cross-sectional area at constant temperature?

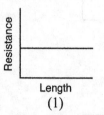
Length
(1)

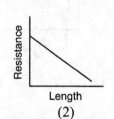
Length
(2)

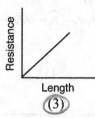
Length
(3)

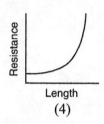
Length
(4)

25. The diagram below represents an electric circuit consisting of a 12-volt battery, a 3.0-ohm resistor, R_1, and a variable resistor, R_2.

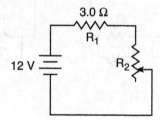

At what value must the variable resistor be set to produce a current of 1.0 ampere through R_1?

(1) 6.0 Ω   (2) 9.0 Ω   (3) 3.0 Ω   (4) 12 Ω

26. The diagram below represents a lamp, a 10-volt battery, and a length of nichrome wire connected in series.

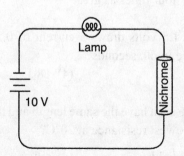

As the temperature of the nichrome is decreased, the brightness of the lamp will

(1) decrease

(2) increase

(3) remain the same

27. If 10. coulombs of charge are transferred through an electric circuit in 5.0 seconds, then the current in the circuit is

(1) 0.50 A   (2) 2.0 A   (3) 15 A   (4) 50. A

28. The diagram below shows the lines of magnetic force between two north magnetic poles.

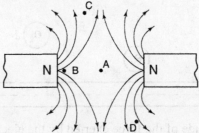

At which point is the magnetic field strength greatest?

(1) A   (2) B   (3) C   (4) D

29. In order to produce a magnetic field, an electric charge must be

(1) stationary   (2) moving   (3) positive   (4) negative

30. Which diagram best represents magnetic flux lines around a bar magnet?

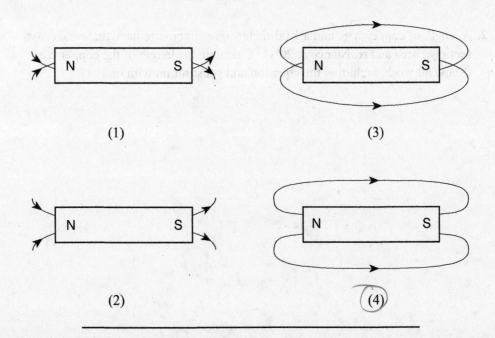

(1)   (3)

(2)   (4)

Part B2

1. The diagram below represents two electrons, e_1 and e_2, located between two oppositely charged parallel plates.

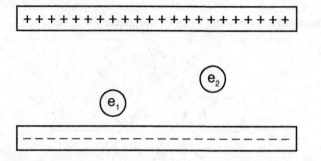

Compare the magnitude of the force exerted by the electric field on e_1 to the magnitude of the force exerted by the electric field on e_2.

_____

_____

2. A length of copper wire and a 1.00-meter-long silver wire have the same cross-sectional area and resistance at 20°C. Calculate the length of the copper wire. (Show all work, including the equation and substitution with units.)

3. Two small identical metal spheres, **A** and **B**, on insulated stands, are each given a charge of $+2.0 \times 10^{-6}$ coulomb. The distance between the spheres is 2.0×10^{-1} meter. Calculate the magnitude of the electrostatic force that the charge on sphere **A** exerts on the charge on sphere **B**. (Show all work, including the equation and substitution with units.)

A 3.0-ohm resistor, an unknown resistor, **R**, and two ammeters, A_1 and A_2, are connected as shown with a 12-volt source. Ammeter A_2 reads a current of 5.0 amperes.

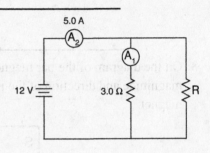

4. Determine the equivalent resistance

of the circuit. _____ Ω

5. Calculate the current measured by ammeter A_1. (Show all work, including equation and substitution with units.)

6. Calculate the resistance of the unknown resistor, **R**. (Show all work, including equation and substitution with units.)

7. A generator produces a 115-volt potential difference and a maximum of 20.0 amperes of current. Calculate the total electrical energy the generator produces operating at maximum capacity for 60. seconds. (Show all work, including the equation and substitution with units.)

---

8. On the diagram of the bar magnet, draw a minimum of four field lines to show the magnitude and direction of the magnetic field in the region surrounding the bar magnet.

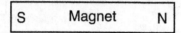

9. An electron is accelerated through a potential difference of 2.5×10^4 volts in the cathode ray tube of a computer monitor. Calculate the work, in joules, done on the electron. (Show all work, including the equation and substitution with units.)

---

10. A student conducted an experiment to determine the resistance of a light bulb. As she applied various potential differences to the bulb, she recorded the voltages and corresponding currents and constructed the graph below.

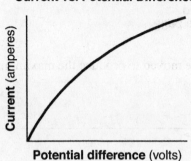

Current vs. Potential Difference

Current (amperes)

Potential difference (volts)

The student concluded that the resistance of the light bulb was not constant. What evidence from the graph supports the student's conclusion?

_____

_____

_____

11. A long copper wire was connected to a voltage source. The voltage was varied and the current through the wire measured, while temperature was held constant. The collected data are represented by the graph below.

Voltage vs. Current

Voltage (V)

20.0

16.0

12.0

8.0

4.0

0.0

0.00   0.20   0.40   0.60   0.80   1.00

Current (A)

Using the graph, determine the resistance of the copper wire.

_____ Ω

12. The diagram below represents a wire conductor, *RS*, positioned perpendicular to a uniform magnetic field directed into the page.

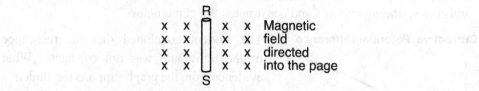

Describe the direction in which the wire could be moved to produce the maximum potential difference across its ends, *R* and *S*.

_____

_____

13. What is the magnitude of the charge, in coulombs, of a lithium nucleus containing three protons and four neutrons?

_____ C

14. A light bulb attached to a 120.-volt source of potential difference draws a current of 1.25 amperes for 35.0 seconds. Calculate how much electrical energy is used by the bulb. (Show all work, including the equation and substitution with units.)

15. The diagram below shows two compasses located near the ends of a bar magnet. The north pole of compass *X* points toward end *A* of the magnet.

On the diagram below, draw the correct orientation of the needle of compass *Y* and label its polarity.

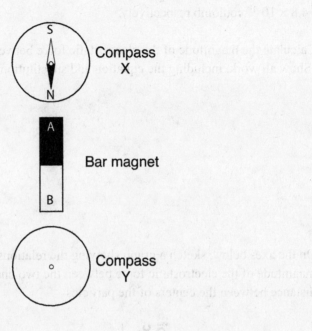

Part C

Base your answers to questions 1 through 3 on the information below:

The centers of two small charged particles are separated by a distance of 1.2×10^{-4} meter. The charges on the particles are $+8.0 \times 10^{-19}$ coulomb, and $+4.8 \times 10^{-19}$ coulomb respectively.

1. Calculate the magnitude of the electrostatic force between these two particles. (Show all work, including the equation and substitution with units.)

2. On the axes below, sketch a graph showing the relationship between the magnitude of the electrostatic force between the two charged particles and the distance between the centers of the particles.

Magnitude of Electrostatic Force

Distance Between Centers

3. On the diagram below draw at least four electric field lines in the region between the two positively charged particles.

8.0×10^{-19} C $(+)$ $(+)$ 4.8×10^{-19} C

Base your answers to questions 4 through 6 on the information below.

A 5.0-ohm resistor, a 10.0-ohm resistor, and a 15.0-ohm resistor are connected in parallel with a battery. The current through the 5.0-ohm resistor is 2.4 amperes.

4. Using the circuit symbols found in the Reference *Tables for Physical Setting/ Physics*, draw a diagram of this electric circuit.

5. Calculate the amount of electrical energy expended in the 5.0-ohm resistor in 2.0 minutes. (Show all work including the equation and substitutions with units.)

6. A 20.0-ohm resistor is added to the circuit in parallel with the other resistors. Describe the effect the addition of this resistor has on the amount of electrical energy expended in the 5.0-ohm resistor in 2.0 minutes.

_____

_____

_____

_____

Base your answers to questions 7 through 9 on the information and diagram below.

A 50.-ohm resistor, an unknown resistor R, a 120-volt source, and an ammeter are connected in a complete circuit. The ammeter reads 0.50 ampere.

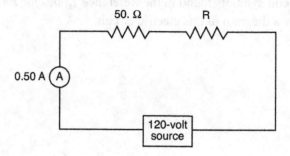

7. Calculate the equivalent resistance of the circuit. (Show all work including the equation and substitution with units.)

8. Determine the resistance of resistor R.

_____ Ω

9. Calculate the power dissipated by the 50.-ohm resistor. (Show all work, including the equation and substitution with units.)

10. The diagram below shows two resistors, R_1 and R_2, connected in parallel in a circuit having a 120-volt power source. Resistor R_1 develops 150 watts and resistor R_2 develops an unknown power. Ammeter A in the circuit reads 0.50 ampere.

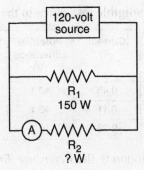

Calculate the amount of charge passing through resistor R_2 in 60. seconds. (Show all work, including the equation and substitution with units.)

_____

Base your answers to questions 11 through 15 on the information and data table below.

Three lamps were connected in a circuit with a battery of constant potential. The current, potential difference, and resistance for each lamp are listed in the data table below. (There is a negligible resistance in the wires and the battery.)

	Current (A)	Potential Difference (V)	Resistance (Ω)
lamp 1	0.45	40.1	89
lamp 2	0.11	40.1	365
lamp 3	0.28	40.1	143

11. Using the circuit symbols found in the *Reference Tables for Physical Setting/ Physics*, draw a circuit showing how the lamps and battery are connected.

12. What is the potential difference supplied by the battery? _____ v

13. Calculate the equivalent resistance of the circuit. (Show all work, including the equation and substitution with units.)

14. If lamp 3 is removed from the circuit, what would be the value of the potential difference across lamp 1 after lamp 3 is removed?

_____ v

15. If lamp 3 is removed from the circuit, what would be the value of the current in lamp 2 after lamp 3 is removed?

_____ A

PART IV: MODERN PHYSICS

UNIT 13
Quantum Theory

Modern Physics emphasizes the *microscopic* properties of matter, rather than the *macroscopic* properties that have been studied thus far. To a typical high school student, the era of focus for modern physics may not seem very modern at all - the early 1900s to the present. Compared to Newtonian Physics, which dates back to the 1600s, it is considered quite modern.

Wave-Particle Duality

Experiments with visible light have shown very clearly that light behaves like a wave. There are certain interactions that only waves can undergo, and light has demonstrated these interactions. The main proponent of the wave theory of light was Christiaan Huygens, a Dutch physicist in the 1600s. These interactions include:

- **Interference:** the combining of two or more waves constructively to produce a wave of higher amplitude (brightness of light), or destructively to produce a wave of lower amplitude.

- **Refraction:** the bending of a wave due to the change in speed as it enters a new medium.

- **the Doppler Effect:** the apparent shift in frequency of a wave as the source and receiver are in relative motion.

- **Polarization:** when transverse waves, such as light, will have all of their vibrations in the same planar orientation.

- **Diffraction:** the spreading of waves past a barrier or opening. Through two openings the light combines in such a way to produce alternating bands of bright and dark regions (Young's double-slit experiment).

As light was further studied, some interactions were observed that did not apply to wave energy, but rather to a particle model of matter. Isaac Newton developed the corpuscular (particle) theory of light, which stated that light was, in fact, a bundle of particles. He could easily explain the reflection of light using this theory. Since light can behave like a wave *and* a particle, the term wave-particle duality was used to describe this behavior.

The one interaction that helped solidify the particle theory of light was the **Photoelectric Effect**. The basis behind the photoelectric effect is that if a specific *photosensitive* metal were to be exposed to certain frequencies of light, that light energy would transmit to the metal and eject electrons (called *photoelectrons*). The only way for light energy to cause these electrons to be ejected, as a cue ball breaks the rack of pool balls in billiards, is for that light to possess particle-like properties. In 1905, Albert Einstein explained the photoelectric effect by stating that light consists of bundles of energy called **photons.** The energy of a photon of light is directly related to its frequency with a constant called **Planck's constant (h):**

$$E_{photon} = hf$$

Christiaan Huygens

Christiaan Huygens was a Dutch mathematician, astronomer, and physicist who pioneered the wave theory of light. In addition, he also discovered Saturn's moon, Titan. Huygens invented the pendulum clock and also derived the formula for centripetal force.

Born: April 14, 1629; The Hague
Died: July 8, 1695; The Hague

Colleges Attended: University of Leiden; College of Orange

Albert Einstein

Albert Einstein was a German born theoretical physicist who is considered to be the father of modern physics. He is the most well-known and influential physicist of the 20th century and developed the theory of relativity. He won the Nobel Prize in Physics in 1921 for explaining the photoelectric effect, and recommended that U.S. government begin research in nuclear fission - the Manhattan Project.

Born: March 14, 1879; Germany
Died: April 18, 1955; United States

Colleges Attended: ETH Zurich; University of Zurich

QUANTUM THEORY

Since the relationship between speed, frequency, and wavelength is known ($v = f\lambda$), the equation on the previous page can be expressed in terms of wavelength:

$$E_{photon} = \frac{hc}{\lambda}$$

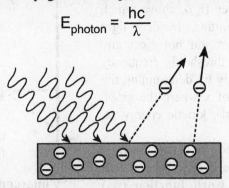

Experiments with the photoelectric effect tested the outcome when, separately, the intensity (amplitude) of the incident light, the frequency (color) of the incident light, and the type of metal used were changed.

- **Effect of intensity:** It was found that increasing the intensity of the incident light caused a larger number of photoelectrons to be emitted from the sample. This is sometimes referred to as the *photocurrent* when part of a circuit. However, there was no effect on the speed (and therefore kinetic energy) of each individual electron.

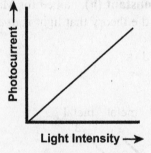
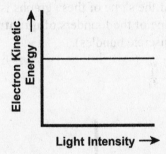

- **Effect of frequency:** It was found that increasing the frequency of the incident light resulted in each individual electron traveling at a faster speed (and therefore having a greater kinetic energy), but there was no effect on the total number of photoelectrons emitted.

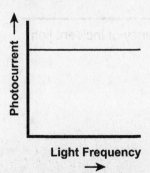
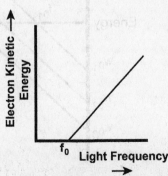

From experimentation, it was determined that there was a minimum frequency (called the **threshold frequency (f_0)**), above which electrons would be emitted. Incident light below this frequency would not eject any photoelectrons. The threshold frequency can be found graphically by determining the x-intercept (the value of x where the graph crosses the x-axis) of the kinetic energy vs. frequency graph.

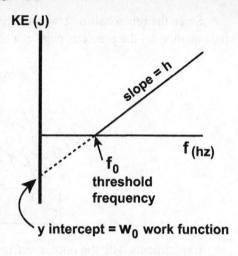

It was also found that extrapolating the graph back to the y-axis proved that the y-intercept is the **work function (w_0)** of the metal. The work function represents the minimum amount of light energy that must fall on the metal in order to eject photoelectrons. The relationship between work function and threshold frequency is:

$$w_0 = hf_0$$

- **Effect of various metals:** By using different photosensitive metals, it was shown that different metals may have different threshold frequencies, but the slope of the KE vs. frequency graphs were all the same. The numerical value of the slope of these graphs is **Planck's Constant (h)**, named for Max Planck, one of the founders of **quantum theory** - the theory that light is *quantized* (in discrete bundles).

$$h = 6.63 \times 10^{-34} \, J \cdot s$$

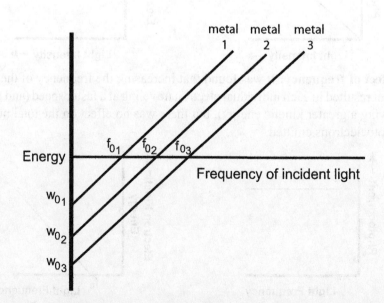

Max Planck

Max Planck was a German theoretical physicist who was the originator of quantum theory. He studied blackbody radiation and sought to improve the efficiency of electric light bulbs. He introduced the Planck postulate - the fact that energy of oscillation in blackbodies is quantized. Planck's Constant (h) is named after him.

Born: April 23, 1858; Germany
Died: October 4, 1942; Germany

College Attended: University of Munich
Won the Nobel Prize in Physics in 1918

The **Photoelectric Equation** connects all of the concepts shown on the previous page to express the kinetic energy of the ejected photoelectrons in terms of the incident energy and the work function of the metal. This equation is actually a statement of *conservation of energy* for the photoelectric effect:

$$KE_{max} = E_{photon} - w_0$$

-or-

$$KE_{max} = hf - hf_0$$

Throughout this time of great investigation and discovery, scientists experimented with a variety of variables in the photoelectric effect. Much of the research at the time was done using visible light, but it was determined that all electromagnetic radiation could be used. This proved that all electromagnetic radiation exhibited both wave and particle properties.

In 1923, American physicist Arthur Compton exposed a photosensitive metal to x-rays. The result was ejected electrons, as expected. At first, it appeared that conservation of energy was violated, as the kinetic energy of the ejected electrons did not equal the incident energy minus the work function. Upon further research, it was determined that the original x-ray was scattered (with less energy as before) along with the electron, thus continuing to prove conservation of energy. Since the electron has mass and velocity, it must then have momentum. This momentum gain must have been from a momentum loss of the x-ray photon in order for conservation of momentum to be true. This led to the idea that photons (massless particles) must have momentum.

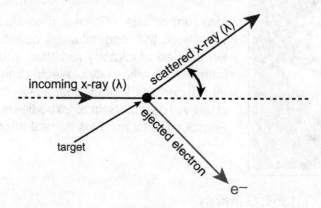

In 1924, Louis deBroglie furthered this idea and postulated that *all matter* must exhibit both particle and wave properties. He especially focused his research on electrons. It was also determined that electrons, typically thought of as particles, can behave like waves when they pass through a double-slit. DeBroglie stated that electrons must then travel in their orbit around the nucleus of an atom in a wavelike pattern.

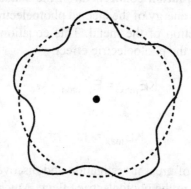

The wavelength or momentum of any matter can be determined by the following equation:

$$\lambda = \frac{h}{p} = \frac{h}{mv}$$

Louis deBroglie stated that all matter travels in a wave. For larger, macroscopic matter this wavelength is unobservable as the wavelength is too small.

Arthur Compton

Arthur Compton was an American physicist most known for his discovery of the Compton Effect - the increasing wavelengths of x-rays during x-ray scattering. This was additional proof of the particle theory of electromagnetic radiation. Compton took over the stagnant American program to develop the atomic bomb. He also invented an improved version of the speed bump for roads, called the Holly Bump, that featured a more elongated shape.

Born: September 10, 1892; United States
Died: March 15, 1962; United States

Nobel Prize in Physics: 1927

Louis de Broglie

Louis de Broglie was a French physicist who won the Nobel Prize in Physics in 1929 for his discovery of the wave nature of electrons. He stated that Einstein's wave and particle duality should "extend to all particles the coexistence of waves and particles." The equation for finding the wavelength of matter waves was determined by de Broglie.

Born: August 15, 1892; France
Died: March 19, 1987; France

College Attended: Sorbonne, University of Paris

Unit 13
Practice Questions

1. Moving electrons are found to exhibit properties of

   (1) particles, only   (3) both particles and waves
   (2) waves, only   (4) neither particles nor waves

2. Which graph best represents the relationship between photon energy and photon frequency?

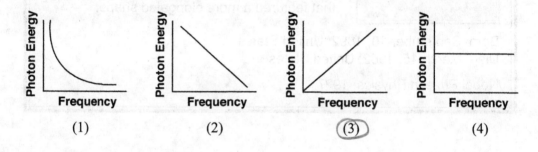

       (1)   (2)   (3)   (4)

---

3. A photon of light traveling through space with a wavelength of 6.0×10^{-7} meter has an energy of

   (1) 4.0×10^{-40} J   (2) 3.3×10^{-19} J   (3) 5.4×10^{10} J   (4) 5.0×10^{14} J

4. All photons in a vacuum have the same

   (1) speed   (2) wavelength   (3) energy   (4) frequency

5. Which phenomenon best supports the theory that matter has a wave nature?

   (1) electron momentum   (3) photon momentum
   (2) electron diffraction   (4) photon diffraction

6. Wave-particle duality is most apparent in analyzing the motion of

   (1) a baseball   (3) a galaxy
   (2) a space shuttle   (4) an electron

7. The energy of a photon is inversely proportional to its

(1) wavelength   (2) speed   (3) frequency   (4) phase

Base your answers to questions 8 and 9 on the data table below. The data table lists the energy and corresponding frequency of five photons.

Photon	Energy (J)	Frequency (Hz)
A	6.63×10^{-15}	1.00×10^{19}
B	1.99×10^{-17}	3.00×10^{16}
C	3.49×10^{-19}	5.26×10^{14}
D	1.33×10^{-20}	2.00×10^{13}
E	6.63×10^{-26}	1.00×10^{8}

8. In which part of the electromagnetic spectrum would photon D be found?

(1) infrared   (2) visible   (3) ultraviolet   (4) x ray

9. The graph below represents the relationship between the energy and the frequency of photons.

Energy vs. Frequency

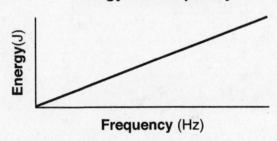

The slope of the graph would be
(1) 6.63×10^{-34} J • s   (3) 1.60×10^{-19} J
(2) 6.67×10^{-11} N • m^2/kg^2 (4) 1.60×10^{-19} C

10. A photon of light carries

(1) energy, but not momentum   (3) both energy and momentum
(2) momentum, but not energy   (4) neither energy nor momentum

NOTES:

UNIT 14
Models of the Atom

There have been many different models of the atom throughout history. Each one sought to improve on the previous, based on more sophisticated measurement and observation techniques. In the early 1800s, John Dalton proposed his atomic theory identifying atoms as indivisible spheres. In 1897, JJ Thomson built upon this idea and identified the electron, and an atom as having both positive and negative parts. There are three later models of the atom, however, that are especially important to the understanding of matter - the **Rutherford model**, the **Bohr model**, and the **electron cloud model**.

The Rutherford Model

In 1909, Ernest Rutherford performed experiments to investigate the structure of the atom. This experiment, the *Geiger-Marsden Experiment* (popularly called the **Rutherford Gold Foil Experiment**) involved bombarding a thin foil of gold with particles from a radioactive source. These particles, called **alpha particles,** came from the radioactive decay of the element radon. The particles were positively charged, and it was later determined that these particles were the nuclei of helium. A helium nucleus consists of 2 protons and 2 neutrons.

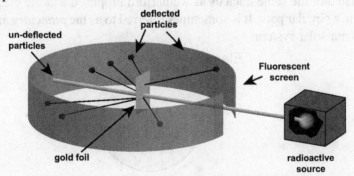

Rutherford expected nearly all of the alpha particles to pass right through the gold foil essentially uninterrupted. JJ Thomson's *plum pudding* model had positive and negative charges spread all throughout the atom, so with the relatively heavy alpha particles, he predicted they would be forced right through the gold. To his surprise, about one of every 8000 alpha particles was deflected in a hyperbolic path, at very large angles, some as large as 180°. Based on the number of particles deflected as compared to the number that passed right through, and the properties of Coulomb's Law (like charges will repel), Rutherford concluded the following:

- most of the atom consists of empty space
- the positive charge of an atom is concentrated in a small dense core, called the *nucleus*
- electrons orbit around the nucleus

When speaking about his findings, Rutherford said:

"It was quite the most incredible event that has ever happened to me in my life. It was almost as incredible as if you fired a 15-inch shell at a piece of tissue paper and it came back and hit you."

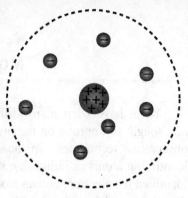

Limitations of the Rutherford Model

Although the Rutherford model was groundbreaking in the fact that the nucleus was identified, there were some limitations that needed further research. The neutron was not accounted for, and the behavior of the electron orbiting the nucleus could not be explained. According to classical mechanics (Newtonian physics), the electron should have accelerated as it circled the nucleus (centripetal acceleration). Because of this acceleration, the atom should emit photons of energy, resulting in the electron spiraling into the nucleus, but it does not.

The Bohr Model

In 1913, Neils Bohr improved on the Rutherford model by attempting to explain the behavior of the electron as it orbits the nucleus. He performed his experiments on the element *hydrogen*, which is now known to consist only of a proton and an electron. His model consisted of the same nucleus as Rutherford proposed and the electron orbiting the nucleus in a circular path. It is sometimes referred to as the *planetary model* because it resembles our solar system.

Electron

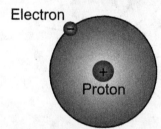

Proton

Bohr's model was a satisfactory model for the simple atom of hydrogen. It was not as accurate a depiction of elements with higher atomic numbers, however. In conducting his research, Bohr studied the **emission spectrum** of hydrogen. To explain his model, Bohr made the following assumptions and conclusions:

- Even though the electron does undergo centripetal acceleration, it does not emit photons of electromagnetic energy.
- The energy of an electron is *quantized* - it exists in discrete, specific amounts. Therefore, electrons can exist at very specific *energy levels* associated with their energy.
- There are a specific number of electrons that can occupy a specific energy level.
- In order for an electron to change energy levels, energy must either be absorbed (to go to a higher level), or emitted (to drop to a lower energy level).

Ernest Rutherford

Ernest Rutherford was a British chemist and physicist, credited as being the father of nuclear physics. He conducted the gold-foil experiment discovering the atomic nucleus. Rutherford also discovered and named the proton. He discovered atomic half-life, for which he was awarded the Nobel Prize in Chemistry in 1908.

Born: August 30, 1871; New Zealand
Died: October 19, 1937; England

Colleges Attended: University of Canterbury, University of Cambridge

Niels Bohr

Niels Bohr was a Danish physicist most known for his planetary model of the atom. He studied atomic structure and the spectra of gases. He worked under JJ Thomson and Ernest Rutherford. Bohr proposed the idea that electrons could drop from a higher-energy orbit to a lower one, in the process emitting a photon of discrete energy. He was awarded the Nobel Prize in Physics in 1922. Bohr also worked on the team of British Physicists working on the Manhattan Project.

Born: October 7, 1885; Denmark
Died: November 18, 1962; Denmark

College Attended: University of Copenhagen

Energy Levels

Neils Bohr studied the energy level changes for the element hydrogen. He found that when the hydrogen atom is given energy, the electron in the atom moves from the *ground state* (or *stationary state*), to a higher energy level, called the *excited state*. This process of raising the energy of atoms is called **excitation**. As the electrons will not remain in this excited state, they emit the same amount of energy that was gained and return back to their ground state. This emission of energy is in the form of photons. These photons that are emitted give the unique **emission spectrum** of an element. The emission spectrum can be used to identify the element as each element has a different spectrum.

One way to represent the energy levels of different elements is to create an **energy level diagram**. The energy level diagrams for hydrogen and mercury, as found in the Reference Tables, are below:

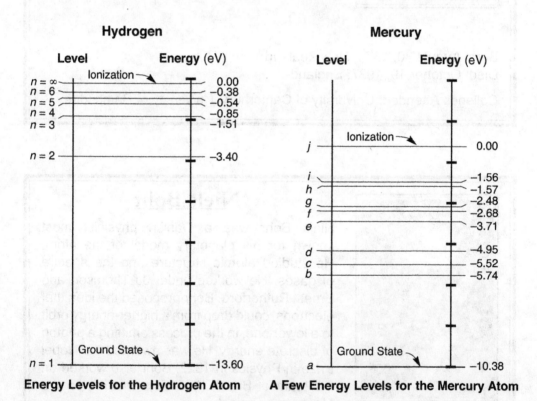

Energy Levels for the Hydrogen Atom   A Few Energy Levels for the Mercury Atom

Note that on the right side of each diagram, the energy associated with a specific level is listed. These energies are in the units of the electron-volt (eV). Recall from earlier units that $1 eV = 1.60 \times 10^{-19} J$. The energy levels for hydrogen are numbered, whereas the levels for mercury are lettered. A diagram like this can be created for all elements.

Ionization occurs when an electron is completely removed from the atom. In order to do this, the electron must be raised from the original energy level to n=∞ (infinity) for hydrogen, or the "j" energy level for mercury. To determine the amount of energy necessary for ionization, simply refer to the energy level diagram for the appropriate element and read the energy associated with the original energy level the electron is in. For example, it would require 3.71 eV of energy to be absorbed by a mercury atom in order to ionize it when an electron is in the "e" energy level initially. For a hydrogen atom in the ground state (n=1), it would require the addition of 13.6 eV of energy in order to ionize it. An electron can absorb more than the amount as determined by the energy level diagram. The extra energy becomes kinetic energy of this ionized electron.

Energy level transitions occur when an electron either absorbs energy to rise to a higher energy level, or release energy to drop to a lower level. It is important to note that the electron must absorb (and subsequently release) the *exact amount of energy* in order to transition levels. An electron can never transition to a point between levels, nor can it absorb more or less energy than it needs to go a level. To determine the exact amount of energy necessary to rise or drop energy levels, the following formula is used:

$$E_{photon} = E_i - E_f$$

where:

E_{photon} = the energy absorbed or emitted (in eV)
E_i = the initial energy level (in eV)
E_f = the final energy level (in eV)

Once the energy absorbed or emitted is determined in eV, it can be converted to Joules, and then put into the equation E_{photon}= hf to determine the frequency of this emitted or absorbed photon, if desired.

An electron can rise several levels during excitation. For example, an electron in a hydrogen atom could rise from the n=1 to the n=4 energy level. To return back to the ground state, it could return directly, releasing a single photon. Or, it could return in stages, such as the n=4 to n=3, then n=3 to n=2, and finally n=2 to n=1. Or, it could return from n=4 to n=2, then from n=2 to n=1. It could also return from n=4 to n=3, then from n=3 to n=1. Each transition releases a photon with a different amount of energy. Drawn pictorially:

Energy level

n = 4	
n = 3	
n = 2	
n = 1	

In the diagram above, you can see that there are 8 total photons that can be emitted, but there are **6 unique photons** that can be emitted, as some of them are repeated and represent the same transition.

There are specific names given to different electron energy level transitions for hydrogen. These transitions are shown on the diagram below:

The **Lyman series** consists of transitions from any higher energy level to the ground state, n=1. The **Balmer series**, which is the transition that gives off the visible spectrum, are transitions from a higher energy level to n=2. The **Paschen series** are transitions to the n=3, the **Brackett series** are transitions to the n=4, and the **Pfund series** are transitions to the n=5.

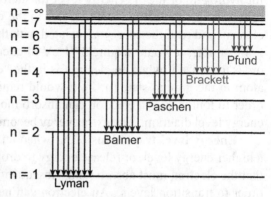

In addition to an emission spectrum, elements can also have an **absorption spectrum**. If light of all colors is sent through a cool gas, the normally continuous spectrum of the white light will have dark lines in it. The dark lines are the frequencies of the photons absorbed by the gas. It was determined that the frequencies of the absorption spectrum exactly match the emission spectrum frequencies when that same gas is excited.

Solved Example Problem

- An atom of hydrogen has an electron in its n = 3 state. The electron spontaneously drops to the n = 1 state (a Lyman transition), emitting a photon.

    a. Calculate the energy of this photon in electron-volts.

    $$E_{photon} = E_i - E_f$$
    $$E_{photon} = (-1.51 \text{ eV}) - (-13.60 \text{ eV})$$
    $$E_{photon} = 12.09 \text{ eV}$$

    the positive value indicates the energy is emitted

    b. Convert this energy into joules.

    $$12.09 \text{ eV}\left(\frac{1.60 \times 10^{-19} \text{J}}{1 \text{ eV}}\right) = 1.93 \times 10^{-18} \text{J}$$

    c. Determine the frequency of this photon.

    $$E_{photon} = hf$$
    $$1.93 \times 10^{-18}\text{J} = (6.63 \times 10^{-34} \text{ J} \cdot \text{s})f$$
    $$f = 2.91 \times 10^{15} \text{ Hz}$$

    d. Use the electromagnetic spectrum chart to determine the photon's type:
    ultraviolet light

The Electron Cloud Model

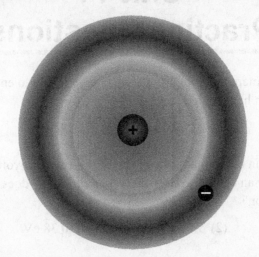

Austrian theoretical physicist Erwin Schrödinger presented a model of the hydrogen atom in 1926 that improved upon the Bohr model. He stated that it is not possible to pinpoint exactly where an electron is at a given moment in time while it is in constant motion. The properties of the nucleus are the same, but there is instead a region of probability where the electron is likely to be located. This region of probability is called the electron *cloud*.

Erwin Schrödinger

Erwin Schrödinger was an Austrian physicist who won the Nobel Prize in Physics in 1933. His award was for the discovery of new productive forms of atomic theory. He was dissatisfied with the quantum condition in Bohr's orbit theory of the atom and sought to explain it in terms of waves, which led to the Schrödinger Wave Equation. He also proposed the Schrödinger's Cat thought experiment, the paradox of the living/dead state of a cat in a box with poison.

Born: August 12, 1887; Austria
Died: January 4, 1961; Austria

College Attended: University of Vienna

Unit 14
Practice Questions

1. An electron in a mercury atom drops from energy level f to energy level c by emitting a photon having an energy of

   (1) 8.20 eV   (2) 5.52 eV   (3) 2.84 eV   (4) 2.68 eV

2. A mercury atom in the ground state absorbs 20.00 electronvolts of energy and is ionized by losing an electron. How much kinetic energy does this electron have after the ionization?

   (1) 6.40 eV   (2) 9.62 eV   (3) 10.38 eV   (4) 13.60 eV

3. A photon having an energy of 9.40 electronvolts strikes a hydrogen atom in the ground state. Why is the photon *not* absorbed by the hydrogen atom?

   (1) The atom's orbital electron is moving too fast.
   (2) The photon striking the atom is moving too fast.
   (3) The photon's energy is too small.
   (4) The photon is being repelled by electrostatic force.

4. Which type of photon is emitted when an electron in a hydrogen atom drops from the $n = 2$ to the $n = 1$ energy level?

   (1) ultraviolet   (3) infrared
   (2) visible light   (4) radio wave

5. A hydrogen atom with an electron initially in the $n = 2$ level is excited further until the electron is in the $n = 4$ level. This energy level change occurs because the atom has

   (1) absorbed a 0.85-eV photon   (3) absorbed a 2.55-eV photon
   (2) emitted a 0.85-eV photon   (4) emitted a 2.55-eV photon

6. The bright-line emission spectrum of an element can best be explained by

   (1) electons transitioning between discrete energy levels in the atoms of that element
   (2) protons acting as both particles and waves
   (3) electrons being located in the nucleus
   (4) protons being dispersed uniformly throughout the atoms of that element

7. The diagram below represents the bright-line spectra of four elements, A, B, C, and D, and the spectrum of an unknown gaseous sample.

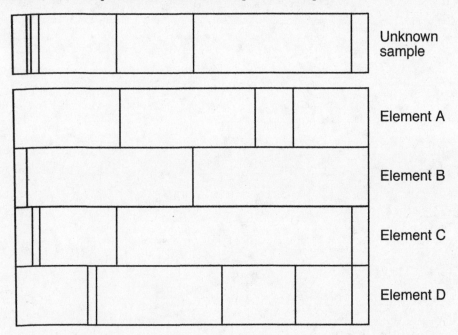

Based on comparisons of these spectra, which two elements are found in the unknown sample?

(1) A and B (2) A and D (3) B and C (4) C and D

8. Excited hydrogen atoms are all in the $n = 3$ state. How many different photon energies could possibly be emitted as these atoms return to the ground state?

(1) 1   (2) 2   (3) 3   (4) 4

9. White light is passed through a cloud of cool hydrogen gas and then examined with a spectroscope. The dark lines observed on a bright background are caused by

(1) the hydrogen emitting all frequencies in white light
(2) the hydrogen absorbing certain frequencies of the white light
(3) diffraction of the white light
(4) constructive interference

10. What is the minimum energy needed to ionize a hydrogen atom in the $n = 2$ energy state?

(1) 13.6 eV   (2) 10.2 eV   (3) 3.40 eV   (4) 1.89 eV

UNIT 15
Elementary Particles and The Standard Model

The Nucleus

The nucleus of an atom contains all of the atom's positive charge. The major particles in the nucleus are the *proton* and the *neutron*. The proton is positively charged (+1 elementary charge, or $+1.60 \times 10^{-19}$ C). The mass of a proton is 1.67×10^{-27} kg. Discovered by James Chadwick in 1932, the neutron has no charge (neutral). The mass of a neutron is 1.67×10^{-27} kg, to three significant figures.

The number of protons in a nucleus is the **atomic number.** The total number of protons and neutrons in a nucleus is the **mass number** of that element. All particles in the nucleus are called **nucleons.** An atomic nucleus can be represented in symbolic form as follows:

$$^{Z}_{A}Sy$$

where:
Sy = the chemical symbol of the element
Z = the mass number (total number of nucleons)
A = the atomic number (total number of protons)

For example, an alpha particle, like the one used in Rutherford's experiments, would be represented like this:

$$^{4}_{2}He$$

This particle has an atomic number of 2. It has a mass number of 4. Therefore, it has 2 protons and 2 neutrons.

A nucleus of carbon, called Carbon-12, would be represented as follows:

$$^{12}_{6}C$$

In addition to carbon in this form, there are other forms of carbon nuclei. A nucleus with the same number of protons (atomic number), but a different number of neutrons (and therefore a different mass number) is called an **isotope**. For example, two isotopes of carbon are Carbon-13 , and Carbon-14 . Carbon actually has 16 different isotopes, but these three are the only ones that are naturally occurring. Carbon-14 is used for *radioactive dating* to determine the age of an organic sample. Many of the elements on the periodic table have naturally occurring isotopes.

The Universal Mass Unit

The mass of subatomic particles is, of course, very small. Much like the Coulomb, which is a unit used to designate a large amount of electrons, there is also a unit to represent a very small amount of mass. This unit is the **Universal Mass Unit (u)**. It can also be referred as the *atomic mass unit (amu)*. It is the standard unit that is used for indicating mass on an atomic or molecular scale. The universal mass unit is defined as 1/12 the mass of a Carbon-12 nucleus. In kilograms, one universal mass unit is equal to $1.66053886 \times 10^{-27}$ kg. The three main subatomic particles' masses, expressed in universal mass units are then:

$$\text{proton mass} = 1.0073 \text{ u}$$
$$\text{neutron mass} = 1.0087 \text{ u}$$
$$\text{electron mass} = 0.00054 \text{ u}$$

As can be observed in the data above, the mass of a neutron is actually a bit higher than the mass of a proton. Expressed in kilograms to three significant figures, however, they are the same.

Nuclear Forces

As the nucleus contains protons and neutrons, there are various forces both holding together these particles, and also forcing them apart. Up to this point, it is known that the **electrostatic force** would cause protons to repel one another. The **gravitational force** would cause them to attract each other. But how do these forces compare? It these were the only two forces in the nucleus, the particles would repel one another as the electrostatic force of repulsion is much stronger than the gravitational force of attraction. There are, however, two other forces responsible for holding the particles of the nucleus together. The **weak nuclear force** is a very short- range force that is responsible for *beta decay* (the nuclear decay of a neutron into a proton, an electron, and another particle called a neutrino). The range of this force is on the order of 10^{-18} m. If two particles are at a distance greater than this, the force goes to zero. The last of the fundamental forces of nature, the **strong nuclear force,** is of slightly longer range than the weak force, 10^{-15} m. This, however, is the strongest of the forces in the nucleus and it is responsible for overcoming forces of repulsion to keep the nucleus intact. When the strong nuclear force is overcome from the addition of energy, *nuclear fission* occurs and large amounts of energy are released.

Binding Energy and Mass Defect

If enough energy is added to an atomic nucleus, it would overcome the strong nuclear force, and the particles would separate. The amount of energy holding the nucleus together (and therefore what must be added to break it apart) is called the **binding energy** of the nucleus. The relative stability of an atomic nucleus is determined by its **binding energy per nucleon**.

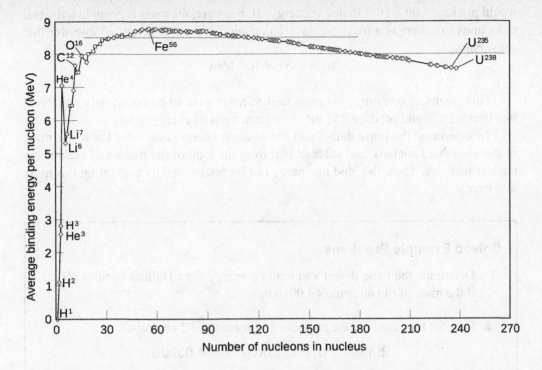

As shown in the graph above, Iron-56 is the most stable nucleus. As the binding energy per nucleon increases, it requires more energy to break it apart, so it is therefore more stable.

It has been determined that the mass of an assembled nucleus is actually less than the total of the masses of the individual particles that make it up. This difference in mass is called the **mass defect** of that nucleus. The difference in mass is associated with the binding energy. As the energy is added to separate the nucleons, the added energy is converted into mass of the nucleons.

This realization has led scientists to redefine the conservation laws. No longer can conservation of mass and conservation of energy be considered separate conservation laws. Now, for nuclear processes, **conservation of mass-energy** as a single system must be considered. Albert Einstein, in his famous equation, relates the conversion between mass and energy:

$$E = mc^2$$

where:
E = energy (in Joules)
m = mass (in kilograms)
c = speed of light in a vacuum (3.00×10^8 m/s)

This equation is used to determine the amount of energy that would be released if a given mass, in kilograms, were completely converted to energy; a process that occurs on the sun, for example. If one kilogram were completely converted into energy, it would produce 9.00×10^{16} Joules of energy. If, however, the mass is given in universal mass units (u), there is a mass-energy relationship from the Reference Tables for that conversion:

$$1u = 9.31 \times 10^2 \text{ MeV}$$

This means, if one universal mass unit of mass were to be completely converted into energy, it would produce 931 mega electron-volts of energy.

To determine the mass defect and the binding energy, one must know the mass of the assembled nucleus, and subtract that from the total of the masses of each of the nuclear particles. Then, the binding energy can be determined by converting this mass into energy.

Solved Example Problems

1. Determine the mass defect and binding energy for a Helium nucleus $\left(^4_2\text{He}\right)$, if the mass of the nucleus is 4.0016 u.

   a. Find the total mass of the nucleons (2 protons and 2 neutrons):

   $$2(1.0073 \text{ u}) + 2(1.0087 \text{ u}) = 4.0320 \text{ u}$$

   b. Subtract the mass of the nucleus from the mass of the nucleons:

   $$4.0320 \text{ u} - 4.0016 \text{ u} = 0.0304 \text{ u}$$

   c. Convert this mass defect into MeV to determine the binding energy:

   $$0.0304 \text{ u} \left(\frac{9.31 \times 10^2 \text{ MeV}}{1\text{u}} \right) = 28.3 \text{ MeV}$$

2. How much energy is released if 5.50 kg of mass were completely converted into energy?

   $$E = mc^2$$
   $$E = (5.50 \text{ kg})(3.00 \times 10^8 \text{ m/s})^2$$
   $$E = 4.95 \times 10^{17} \text{ J}$$

Antimatter

The study of cosmic radiation led scientists to the discovery of positive electrons, a type of **antimatter**. Antimatter has the same properties of regular matter such as mass, except opposite charge, and some other properties. For example, an *antiproton* has the same mass as a proton, but it is negatively charged. An antielectron, called a *positron*, is a positively charged electron. An *antineutron* is neutral, like a regular neutron, but there are other characteristics that differ from a neutron. If matter and antimatter collide, their mass will annihilate into energy.

Solved Example Problem

- Determine the total amount of energy released when an electron and a positron collide and annihilate each other.

$$E = mc^2$$
$$E = (2 \cdot 9.11 \times 10^{-31}\,kg)(3.00 \times 10^8\,m/s)^2$$
$$E = 1.64 \times 10^{-13}\,J$$

Elementary Particles and The Standard Model

In addition to the proton, neutron, and electron, the existence of several hundred additional particles has been discovered since the second half of the 20th Century. Many of these particles are unstable, but they have been found to exist. In order to categorize these particles, the **Standard Model** of subatomic particles has been developed to classify matter.

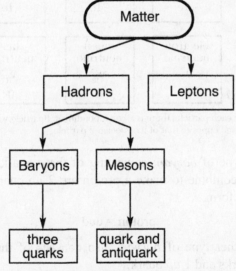

The chart above, from the Reference Tables, shows the basic classification of all matter known to exist today. All matter can be classified as either a *hadron* or a *lepton*. As can be observed by looking at the hadron and lepton categorization above, hadrons are made of quarks, and leptons are not. Since there are no branches below the lepton, it is believed at this point that leptons can't be broken down to smaller particles. Hadrons, however, can be broken down to smaller particles called **quarks**. A quark is a fundamental particle that combines to form baryons and mesons. Quarks have never been observed by themselves; they are always combined with other quarks or an antiquark. If a particle consists of three quarks, that particle is called a *baryon*. If the particle consists of a quark and an antiquark combined, then that particle is called a *meson*. The properties of the quarks and leptons are in the chart on the following page, from the Reference Tables:

Particles of the Standard Model

Quarks

	up	charm	top
Name	up	charm	top
Symbol	u	c	t
Charge	$+\frac{2}{3}\,e$	$+\frac{2}{3}\,e$	$+\frac{2}{3}\,e$

down	strange	bottom
down	strange	bottom
d	s	b
$-\frac{1}{3}\,e$	$-\frac{1}{3}\,e$	$-\frac{1}{3}\,e$

Leptons

electron	muon	tau
electron	muon	tau
e	μ	τ
$-1e$	$-1e$	$-1e$

electron neutrino	muon neutrino	tau neutrino
electron neutrino	muon neutrino	tau neutrino
ν_e	ν_μ	ν_τ
0	0	0

Note: For each particle, there is a corresponding antiparticle with a charge opposite that of its associated particle.

The proton is a type of *baryon*; consisting of three quarks in combination. The particular quarks that combine to make a proton are 2 *up* quarks and 1 *down* quark. Expressed in symbolic form:

$$proton = uud$$

The neutron, another type of *baryon,* also consists of three quarks. A neutron consists of 2 *down* quarks and 1 *up* quark:

$$neutron = ddu$$

There are many other baryons that have been observed. Because they consist of three quarks, their total charges can only be -1, 0, +1, or +2 elementary charges. Although other quark combinations may produce total charges of +1 or 0, those particular combinations are not protons or neutrons. Only the combinations of quarks listed above are protons and neutrons.

Mesons consist of one quark and one antiquark. Recall that antimatter has similar properties to regular matter, except the charge and certain other properties are opposite. The representation of an antiquark is to put a bar over the symbol. For example, an anti-up quark (which is NOT a down quark) would be represented as $\bar{u}$. An example of a meson is $c\bar{s}$, which has a total charge of +1 elementary charge.

Unit 15
Practice Questions

1. Which fundamental force is primarily responsible for the attraction between protons and electrons?

   (1) strong
   (2) weak
   (3) gravitational
   (4) electromagnetic

2. The total conversion of 1.00 kilogram of the Sun's mass into energy yields

   (1) 9.31×10^2 MeV
   (2) 8.38×10^{19} MeV
   (3) 3.00×10^8 J
   (4) 9.00×10^{16} J

Base your answers to questions 3 and 4 on the table below, which shows data about various subatomic particles.

Subatomic Particle Table

Symbol	Name	Quark Content	Electric Charge	Mass (GeV/c²)
p	proton	uud	+1	0.938
p̄	antiproton	ūūd̄	−1	0.938
n	neutron	udd	0	0.940
λ	lambda	uds	0	1.116
Ω⁻	omega	sss	−1	1.672

3. Which particle listed on the table has the opposite charge of, and is more massive than a proton?

   (1) antiproton   (2) neutron   (3) antimatter   (4) omega

4. All the particles listed on the table are classified as

   (1) mesons   (2) hadrons   (3) antimatter   (4) leptons

5. Baryons may have charges of

(1) +1e and $+\frac{4}{3}$e

(2) +2e and +3e

(3) -1e and +1e

(4) -2e and $-\frac{2}{3}$e

6. What fundamental force holds quarks together to form particles such as protons and neutrons?

(1) electromagnetic force

(2) gravitational force

(3) strong force

(4) weak force

7. A tritium nucleus is formed by combining two neutrons and a proton. The mass of this nucleus is 9.106×10^{-3} universal mass unit less than the combined mass of the particles from which it is formed. Approximately how much energy is released when this nuleus is formed?

(1) 8.48×10^{-2} MeV

(2) 2.73 MeV

(3) 8.48 MeV

(4) 273 MeV

8. A lithium atom consists of 3 protons, 4 neutrons, and 3 electrons. This atom contains a total of

(1) 9 quarks and 7 leptons

(2) 12 quarks and 6 leptons

(3) 14 quarks and 3 leptons

(4) 21 quarks and 3 leptons

9. According to the Standard Model of Particle Physics, a meson is composed of

(1) a quark and a muon neutrino

(2) a quark and an antiquark

(3) three quarks

(4) a lepton and an antilepton

10. The tau neutrino, the muon neutrino, and the electron neutrino are all

(1) leptons

(2) hadrons

(3) baryons

(4) mesons

11. According to the Standard Model, a proton is constructed of two up quarks and one down quark (*uud*) and a neutron is constructed of one up quark and two down quarks (*udd*). During beta decay, a neutron decays into a proton, an electron, and an electron antineutrino. During this process there is a conversion of a

(1) *u* quark to a *d* quark

(2) *d* quark to a meson

(3) baryon to another baryon

(4) lepton to another lepton

12. The energy equivalent of the rest mass of an electron is approximately

(1) 5.1×10^5 J

(2) 8.2×10^{-14} J

(3) 2.7×10^{-22} J

(4) 8.5×10^{-28} J

13. The charge-to-mass ratio of an electron is

(1) 5.69×10^{-12} C/kg   (3) 1.76×10^{11} C/kg

(2) 1.76×10^{-11} C/kg   (4) 5.69×10^{12} C/kg

14. If a deuterium nucleus has a mass of 1.53×10^{-3} universal mass units less than its components, this mass represents an energy of

(1) 1.38 MeV   (2) 1.42 MeV   (3) 1.53 MeV   (4) 3.16 MeV

Base your answer to question 15 on the cartoon below and your knowledge of physics

15. In the cartoon, Einstein is contemplating the equation for the principle that

(1) the fundamental source of all energy is the conversion of mass into energy

(2) energy is emitted or absorbed in discrete packets called photons

(3) mass always travels at the speed of light in a vacuum

(4) the energy of a photon is proportional to its frequency

NOTES:

UNITS 13-15 Cumulative Review
Part A and B1

1. Light demonstrates the characteristics of
   (1) particles, only
   (2) waves, only
   (3) both particles and waves
   (4) neither particles nor waves

2. The energy produced by the complete conversion of 2.0×10^{-5} kilogram of mass into energy is
   (1) 1.8 TJ
   (2) 6.0 GJ
   (3) 1.8 MJ
   (4) 6.0 kJ

3. The charge of an antistrange quark is approximately
   (1) $+5.33 \times 10^{-20}$ C
   (2) -5.33×10^{-20} C
   (3) $+5.33 \times 10^{20}$ C
   (4) -5.33×10^{20} C

4. Light of wavelength 5.0×10^{-7} meter consists of photons having an energy of
   (1) 1.1×10^{-48} J
   (2) 1.3×10^{-27} J
   (3) 4.0×10^{-19} J
   (4) 1.7×10^{-5} J

5. Which statement is true of the strong nuclear force?
   (1) It acts over very great distances
   (2) It holds protons and neutrons together.
   (3) It is much weaker than gravitational forces.
   (4) It repels neutral charges.

6. A meson may *not* have a charge of
   (1) + 1e
   (2) +2e
   (3) 0e
   (4) −1e

7. How much energy is required to move an electron in a mercury atom from the ground state to energy level h?
   (1) 1.57 eV
   (2) 8.81 eV
   (3) 10.38 eV
   (4) 11.95 eV

8. Which combination of quarks could produce a neutral baryon?
   (1) *cdt*
   (2) *cts*
   (3) *cdb*
   (4) *cdu*

9. The force that holds protons and neutrons together is known as the
   (1) gravitational force
   (2) strong force
   (3) magnetic force
   (4) electrostatic force

10. The energy equivalent of 5.0×10^{-3} kilogram is
(1) 8.0×10^5 J   (3) 4.5×10^{14} J
(2) 1.5×10^6 J   (4) 3.0×10^{19} J

11. An electron in a mercury atom drops from energy level i to the ground state by emitting a single photon. This photon has an energy of
(1) 1.56 eV   (2) 8.82 eV   (3) 10.38 eV   (4) 11.94 eV

12. Compared to a photon of red light, a photon of blue light has a
(1) greater energy   (3) smaller momentum
(2) longer wavelength   (4) lower frequency

13. Protons and neutrons and examples of
(1) positrons   (2) baryons   (3) mesons   (4) quarks

14. A photon of which electromagnetic radiation has the most energy?
(1) ultraviolet   (3) infrared
(2) x ray   (4) microwave

15. After electrons in hydrogen atoms are excited to the $n = 3$ energy state, how many frequencies of radiation can be emitted as the electrons return to the ground state?
(1) 1   (2) 2   (3) 3   (4) 4

Part B2

1. Calculate the wavelength of a photon having 3.26×10^{-19} joule of energy. [Show all work, including the equation and substitution with units.]

$$E = hf$$
$$3.26 \times 10^{-19} \text{ J} = (6.63 \times 10^{-34} \text{ J} \cdot \text{s}) f$$
$$f = 4.92 \times 10^{14} \text{ Hz}$$

2. If a proton were to combine with an antiproton, they would annihilate each other and become energy. Calculate the amount of energy that would be released by this annihilation. [Show all work, including the equation and substitution with units.]

$$E = mc^2$$
$$E = 2(1.67 \times 10^{-27} kg)(3 \times 10^8 \, m/s)^2$$
$$\boxed{E = 3.01 \times 10^{-10} \, J}$$

3. After uranium nucleus emits an alpha particle, the total mass of the new nucleus and the alpha particle is less than the mass of the original uranium nucleus. Explain what happens to the missing mass.

This is because of mass defect, which states that all the masses of the particles making up the atom are greater than the mass of the atom because the mass converts to energy

4. How much energy, in megaelectronvolts, is produced when 0.250 universal mass unit of matter is completely converted into energy?

_____233_____ MeV

5. What are the sign and charge, in coulombs, of an antiproton?

_____−1.6×10⁻¹⁹_____ C

6. A lambda particle consists of an up, a down, and a strange quark. What is the charge of a lambda particle in elementary charges?

_____0_____ e

Base your answers to questions 7 and 8 on the information and equation below.

During the process of beta (β^-) emission, a neutron in the nucleus of an atom is converted into a proton, an electron, an electron antineutrino, and energy.

$$\text{neutron} \rightarrow \text{proton} + \text{electron} +$$
$$\text{electron antineutrino} + \text{energy}$$

7. Based on conservation laws, how does the mass of the neutron compare to the mass of the proton?

Mass of neutron is greater because it can convert into a proton and mass defect makes the proton less

8. Since charge must be conserved in the reaction shown, what charge must an electron antineutrino carry?

0

Base your answers to questions 9 and 10 on the information below.

When an electron and its antiparticle (positron) combine, they annihilate each other and become energy in the form of gamma rays.

9. The positron has the same mass as the electron. Calculate how many joules of energy are released when they annihilate. [Show all work, including the equation and substitution with units.]

$$E = mc^2$$
$$E = 2(9.11 \times 10^{-31} \text{ kg})(3 \times 10^8 \text{ m/s})^2$$
$$E = 1.64 \times 10^{-13} \text{ J}$$

10. What conservation law prevents this from happening with two electrons?

Conservation of charge

Part C

Base your answers to questions 1 through 3 on the passage below:

For years, theoretical physicists have been refining a mathematical method called lattice quantum chromodynamics to enable them to predict the masses of particles consisting of various combinations of quarks and antiquarks. They recently used the theory to calculate the mass of the rare B_c particle, consisting of a charm quark and a bottom antiquark. The predicted mass of the B_c particle was about six times the mass of a proton.

Shortly after the prediction was made, physicists working at the Fermi National Accelerator Laboratory, Fermilab, were able to measure the mass of the B_c particle experimentally and found it to agree with the theoretical prediction to within a few tenths of a percent. In the experiment, the physicists sent beams of protons and antiprotons moving at 99.999% the speed of light in opposite directions around a ring 1.0 kilometer in radius. The protons and antiprotons were kept in their circular paths by powerful electromagnets. When the protons and antiprotons collided, their energy produced numerous new particles, including the elusive B_c.

These results indicate that lattice quantum chromodynamics is a powerful tool not only for confirming the masses of existing particles, but also for predicting the masses of particles that have yet to be discovered in the laboratory.

1. Identify the class of matter to which the B_c particle belongs.

_____

2. Determine both the sign and the magnitude of the charge of the B_c particle in elementary charges.

_____

3. Explain how it is possible for a colliding proton and antiproton to produce a particle with six times the mass of either.

_____

_____

_____

Base your answers to questions 4 and 5 on the information and the data table below.

In the first nuclear reaction using a particle accelerator, accelerated protons bombarded lithium atoms, producing alpha particles and energy. The energy resulted from the conversion of mass into energy. The reaction can be written as shown below.

$$_1^1H + {}_3^7Li \rightarrow {}_2^4He + {}_2^4He + \text{energy}$$

Data Table

Particle	Symbol	Mass (u)
proton	$_1^1H$	1.007 83
lithium atom	$_3^7Li$	7.016 00
alpha particle	$_2^4He$	4.002 60

4. Determine the difference between the total mass of a proton plus a lithium atom $_1^1H + {}_3^7Li$, and the total mass of two alpha particles, $_2^4He + {}_2^4He$, in universal mass units.

_____ u

5. Determine the energy in megaelectronvolts produced in the reaction of a proton with a lithium atom.

_____ MeV

_____

Base your answers to questions 6 through 8 on the information below.

A photon with a frequency of 5.02×10^{14} hertz is absorbed by an excited hydrogen atom. This causes the electron to be ejected from the atom, forming an ion.

6. Calculate the energy of this photon in joules. [Show all work, including the equation and substitution with units.]

$$E = hf$$
$$E = (6.63 \times 10^{-34} \ J \cdot s)(5.02 \times 10^{14} \ Hz)$$
$$E = 3.33 \times 10^{-19} \ J$$

7. Determine the energy of this photon in electronvolts.

_____2.08_____ eV

8. What is the number of the *lowest* energy level (closest to the ground state) of a hydrogen atom that contains an electron that would be ejected by the absorption of this photon?

$n =$ _____3_____

Base your answers to questions 9 through 12 on the Energy Level Diagram for Hydrogen in the Reference Tables for Physical Setting /Physics.

9. Determine the energy, in electronvolts, of a photon emitted by an electron as it moves from the $n = 6$ to the $n = 2$ energy level in a hydrogen atom.

   _____3.02_____ eV

10. Convert the energy of the photon to joules.

   _____4.83×10^{-19}_____ J

11. Calculate the frequency of the emitted photon. [Show all work, including the equation and substitution with units.]

   $E = hf$

   $4.83 \times 10^{-19} J = (6.63 \times 10^{-34} J \cdot s) f$

   $f = 7.29 \times 10^{14} Hz$

12. Is this the only energy and/or frequency that an electron in the $n = 6$ energy level of a hydrogen atom could emit? Explain your answer.

   No, because the electron could go down

   to any energy level below n=6

Base your answers to question 13 and 14 on the information below.

Louis de Broglie extended the idea of wave-particle duality to all of nature with his matter-wave equation, $\lambda = \dfrac{h}{mv}$, where λ is the particle's wavelength, m is its mass, v is its velocity, and h is Planck's constant.

13. Using this equation, calculate the de Broglie wavelength of a helium nucleus (mass = 6.7×10^{-27} kg) moving with a speed of 2.0×10^6 meters per second. [Show all work, including the equation and substitution with units.]

14. The wavelength of this particle is of the same order of magnitude as which type of electromagnetic radiation?

_____

_____

NOTES:

PHYSICS FUNDAMENTALS

Unit 1
Mathematics and Vectors: A Toolkit for Physics

√

1. The meter is the SI unit of length. It is a fundamental unit.

2. The second is the SI unit of time. It is a fundamental unit.

3. The kilogram is the SI unit of mass. It is a fundamental unit.

4. The Kelvin is the SI unit of temperature. It is a fundamental unit.

5. The Ampere is the SI unit of current. It is a fundamental unit.

6. Be able to convert between units in the SI system of measurement (dimensional analysis).

7. Be able to distinguish between mass and weight.

8. Distance is a scalar quantity that represents the length of a path from one point to another.

9. Displacement is a vector quantity that represents the length and direction of a straight-line path from one point to another between which motion of an object has taken place.

10. Speed is a scalar quantity that represents the magnitude of the velocity.

11. Velocity is a vector quantity that represents the time-rate of change of displacement.

12. A force is a vector quantity that may be defined as a push or pull.

13. The resultant of two or more concurrent vectors acting on a point is the single vector producing the same effect.

14. The equilibrant is the force that is equal to and opposite of the resultant.

15. Equilibrium is when the vector sum of the concurrent forces acting on an object is zero.

16. A single vector may be resolved into an unlimited number of components.

17. Be able to add vectors from tip to tail, and find the magnitude and direction of the answer (resultant).

18. Be able to resolve a vector into its vertical and horizontal components.

19. The maximum resultant occurs when vectors act at $0°$ (in the same direction). The minimum resultant occurs at $180°$ (in opposite directions).

Unit 2
Kinematics: Constant and Changing Velocities

√

- [ ] 1. **Kinematics** deals with the mathematical methods of describing motion without regard to the forces that produce it.

- [ ] 2. **Speed** is a scalar quantity telling us how far an object travels during every unit of time. **Velocity** is a vector quantity describing both the speed of an object and its direction. The SI unit of speed and velocity is m/s.

- [ ] 3. Be able to distinguish between **average speed** and **instantaneous speed**.

- [ ] 4. In **uniform motion**, both the speed and direction of a moving body remain the same. Motion with changing velocity is called **accelerated motion**.

- [ ] 5. **Acceleration** is a vector quantity that represents the time-rate of change in velocity of an object. The SI unit of acceleration is m/s^2.

- [ ] 6. Be able to recognize, interpret, and use graphs of distance vs. time, velocity vs. time, and acceleration vs. time.

- [ ] 7. The slope of distance vs. time represents velocity, and that of velocity vs. time represents acceleration.

- [ ] 8. If the speed is changing, the distance-time graph is curved, and the slope of the tangent to the distance-time curve at any point represents the instantaneous speed at that point.

- [ ] 9. The area under a velocity vs. time graph represents displacement.

- [ ] 10. Freely falling objects may be considered examples of objects with uniformly accelerated motion.

- [ ] 11. A freely falling body in a vacuum accelerates downward at a rate of $9.81 m/s^2$ (g).

- [ ] 12. Be able to solve problems using the following equations:

$$v = \frac{d}{t}$$

$$v_f = v_1 + at$$

$$d = v_i t + \frac{1}{2}at^2$$

$$a = \frac{\Delta v}{t}$$

$$v_f^2 = v_i^2 + 2ad$$

Unit 3
Dynamics: Forces and Newton's Laws

√

1. Know Newton's Three Laws of Motion, which describe how forces control motion, and understand their limitations.

2. **Statics** deals with the relation between forces acting on an object at rest.

3. An object in **static equilibrium** remains at rest or moves with constant velocity.

4. Dynamics deals with the relation between the forces acting on an object and the resulting change in motion.

5. The **Newton** is the **SI unit** of **force**. It will impart to a mass of one kilogram an acceleration of one meter per second per second. It is a derived unit ($N = kg\ m/s^2$).

6. Be able to define **inertia** and cite examples of the Law of Inertia.

7. The **weight** of an object is the net gravitational force on the object. It is directly proportional to the object's mass. Weight is a **vector quantity**.

8. The slope of a force versus acceleration graph equals the mass of the object.

9. The slope of a weight versus mass graph is equal to g.

10. The unbalanced force (net force) is the vector sum of all the forces acting on the object.

11. Be able to define the following terms: Friction, Coefficient of Friction, Static Friction, Kinetic Friction, Rolling Friction, and Fluid Friction.

12. Know how to draw Free-Body Diagrams for various physical situations.

13. Any two objects, whose dimensions are small in comparison to the distance between them, attract each other with a gravitational force that is directly proportional to the product of their masses and inversely proportional to the square of the distance between them.

14. The magnitude of the strength of a gravitational field at any point is the force per unit mass at that point in the gravitational field. An object can push or pull you without touching you.

15. Be able to solve problems using the following equation.

$$a = \frac{F_{net}}{m} \qquad\qquad F_f = \mu\, F_N$$

$$F_g = \frac{Gm_1 m_2}{r^2} \qquad\qquad g = \frac{F_g}{m}$$

Unit 4
Motion in Two Dimensions: Circles and Projectiles

√

☐ 1. The horizontal and vertical components of a projectile's motion are independent of each other.

☐ 2. Circular motion is composed of two components: a velocity tangent to the circle and an acceleration directed toward the center of a path.

☐ 3. The acceleration directed toward the center is called Centripetal Acceleration and the force that causes it is Centripetal Force. Both are vector quantities.

☐ 4. Be able to solve problems finding the range and maximum height of projectiles.

☐ 5. Be able to solve problems using the following equations:

$$F_c = \frac{mv^2}{r} \qquad v_{iy} = v_i \sin\theta$$

$$a_c = \frac{v^2}{r} \qquad v_{ix} = v_i \cos\theta$$

Unit 5
Momentum

√

☐ 1. **Momentum** is a vector quantity, with a magnitude equal to the product of an object's mass and velocity. The direction is determined by the direction of the velocity. The SI unit for momentum is kg m/s.

☐ 2. **Impulse** is a vector quantity with a magnitude equal to the product of the unbalanced force and the time the force acts. Impulse equals the change in momentum and is in the same direction as the acting force. N • s is the SI unit for impulse.

☐ 3. Know the law of conservation of momentum and be able to apply it to collisions.

☐ 4. Be able to solve problems using the following equations:

$$p = mv \qquad\qquad P_{before} = P_{after}$$

$$J = Ft = \Delta p$$

Unit 6
Energy

☐ 1. When work is done on or by a system the total energy of the system is changed. Energy is needed to do the work.

☐ 2. Work is done on an object when a force displaces the object. Work is a scalar quantity.

☐ 3. Energy is transferred when work is done. Energy is a scalar quantity.

☐ 4. The Joule is the unit of work. It is the work done when a force of one newton acts through a distance of one meter. The joule is also the unit of energy. Joule = Newton meter $(J = N \cdot m)$.

☐ 5. Be able to define and work with potential energy and kinetic energy.

☐ 6. Be able to work with and define conservation of energy, and the relationship between work, potential energy and kinetic energy.

☐ 7. Power is the time-rate of change of doing work. It is a scalar quantity.

☐ 8. The Watt is the unit of power. It is equal to one joule per second $(W = \dfrac{J}{s} = \dfrac{k \cdot gm^2}{s^3})$.

☐ 9. Potential Energy is the energy an object has because of its position or condition. The change in potential energy is equal to the work required to bring the object to that position. Potential energy is a scalar quantity.

☐ 10. Gravitational Potential Energy: if work is done on an object against gravitational force, there is an increase in the gravitational potential energy of the system.

☐ 11. Elastic Potential Energy: when work is done on a spring by compressing or stretching it, potential energy is stored in the spring.

☐ 12. The spring constant is the ratio of the force stretching a spring to the amount the spring is stretched. The unit of measure is newton per meter (N/m).

☐ 13. The slope of a force vs. displacement graph is the spring constant, k. The newton per meter (N/m) is the SI unit for the spring constant.

☐ 14. Kinetic Energy is the energy an object has because of its motion.

☐ 15. Conservation of energy: in any transfer of energy among objects in a closed system, the total energy of the system remains constant.

☐ 16. Be able to solve problems using the following equations:

$$W = Fd = DE_T \qquad P = \frac{W}{t} = \frac{Fd}{t} = Fv \qquad E_T = PE + KE + Q$$

$$\Delta PE = mg\Delta h \qquad F_S = kx$$

$$KE = \frac{1}{2}mv^2 \qquad PE_S = \frac{1}{2}kx^2$$

Unit 7
Waves, Sound and Light Properties

√

☐ **1.** Wave motion transfers energy from one point to another with no transfer of matter between the points.

☐ **2.** A **wave** is a vibratory disturbance that is propagated from a source.

☐ **3.** A wave may be classified as a pulse or a periodic wave.

☐ **4.** A **pulse** is a single vibratory disturbance, which moves from point to point. A pulse may be either a **crest** (upward disturbance) or a **trough** (downward disturbance).

☐ **5.** A **periodic wave** is a series of regular disturbances.

☐ **6.** In a uniform medium, a pulse has a constant speed. The speed depends upon the medium.

☐ **7.** When a pulse reaches a boundary with a different medium, part of the pulse will be **reflected** at the boundary and part will be **transmitted** and/or **absorbed** through the second medium.

☐ **8.** Two simple types of waves are longitudinal and transverse.

☐ **9.** In **longitudinal waves** the displacement of the medium is parallel to the direction of travel of the wave.

☐ **10.** In **transverse waves** the disturbance is at right angles to the direction of travel of the wave.

☐ **11.** A transverse wave is **polarized** when the disturbance is in a single plane. Longitudinal waves cannot be polarized.

☐ **12.** Be able to define the following wave characteristics:

    a. Frequency - the number of cycles per unit time made by a vibrating object. It is determined by the source.

    b. Period - the time for one complete cycle or oscillation. The time required for a single wavelength to pass a given point.

    c. Amplitude - the magnitude of the maximum displacement of a vibrating particle from its equilibrium position. It is a measure of the energy of the wave.

    d. Phase - the stage within each oscillation, both position and direction of movement.

    e. Wavelength (λ) - the distance between two consecutive points of a wave moving in the same direction with the same displacement (i.e. in phase). For a transverse wave one complete crest and one complete trough.

    f. Speed - for a wave, the product of frequency and wavelength. Determined by the medium in which the wave is moving.

√

☐ **13.** A **dispersive medium** is one in which the speed of a wave depends on its frequency.

☐ **14.** The **Doppler effect** is the variation in an observed frequency when there is relative motion between source and receiver

   a. There is an increase in observed frequency when the distance between source and receiver is decreasing.

   b. There is a decrease in observed frequency when distance between source and receiver is increasing.

☐ **15.** A **wave front** is the locus of adjacent points of the wave that are in phase. Every point on a wave front may be considered a source of wavelets with the same speed.

☐ **16.** Know the following wave phenomena:
   a. **Reflection** - the rebounding of waves from the surface of a new medium or barrier.

   b. **Refraction** - the bending of a wave as it enters at an oblique angle a medium in which its speed changes.

   c. **Diffraction** - the spreading of waves around, or through an opening in, an obstacle and into the region behind it.

   d. **Interference** - the mutual reinforcement in some places, and the cancellation in others, of the effects of two superimposed waves.

   e. **Superposition** - the process whereby two or more waves combine their effects when passing through the same parts of a medium at the same time. The displacements of the waves add algebraically.

   f. **Standing Waves**- stationary wave patterns formed in a medium when two sets of waves of equal wavelength and amplitude pass through the medium in opposite directions.

   g. **Resonance** - the effect produced when a vibrating object is forced to vibrate at one of its natural frequencies (sympathetic vibration).

☐ **17.** Be able to solve problems using the following equations:

$$T = \frac{1}{f} \qquad\qquad v = f\lambda$$

Unit 8
Reflection and Refraction

☐ 1. When light is reflected from a surface, the angle of incidence is equal to the angle of reflection **(Law of Reflection)**.

☐ 2. The image formed by a plane reflecting surface is virtual, upright, and the same size as the object; object and image distances are equal.

☐ 3. Be able to recognize **incident ray, reflected ray**, and **normal**.

☐ 4. **Regular reflection** is reflection produced by polished surfaces, usually producing an image of the source.

☐ 5. **Diffuse Reflection** is the scattering of light caused by reflection from irregular surfaces.

☐ 6. The ratio of the sine of the angle of incidence to the sine of the angle of refraction is a constant called the relative **index of refraction**.

☐ 7. Be able to recognize the **refracted ray**.

☐ 8. Know the qualitative law of refraction.

☐ 9. The **index of refraction** of a medium is the ratio of the speed of light in a vacuum to its speed in the material medium.

☐ 10. The **critical angle** is the angle of incidence for which the angle of refraction is 90°.

☐ 11. **Total internal reflection** occurs when light is incident on a surface at an angle greater than the critical angle.

☐ 12. Be able to solve problems using the following equations:

$$\theta_i = \theta_r \qquad n = \frac{c}{v} \qquad \sin\theta_c = \frac{1}{n}$$

$$n_1 \sin\theta_1 = n_2 \sin\theta_2$$

$$\frac{n_2}{n_1} = \frac{v_1}{v_2} = \frac{\lambda_1}{\lambda_2}$$

Unit 9
Diffraction and Interference

√

☐ 1. The separation of light into its component colors from a device such as a prism is called **dispersion**.

☐ 2. Sources that produce waves with a constant phase relationship are said to be **coherent**.

☐ 3. Diffraction is the apparent bending of waves around small obstacles and the spreading out of waves past small openings

☐ 4. **Interference** is a phenomenon in which two waves superimpose to form a resultant wave of greater or lower amplitude.

☐ 5. **Superposition** refers to two (or more) waves or pulses traveling through the same medium at the same time

☐ 6. Maximum **constructive interference** occurs at points where the two waves or pulses are in phase.

☐ 7. Maximum **destructive interference** occurs at points where the phase difference is 180°.

☐ 8. Light from a point or line source is diffracted and produces interference patterns when passing through a **single slit**. The width of the central maximum varies directly to the wavelength and inversely to the width of the slit. The central maximum is wider than the successive maxima.

☐ 9. **Double slit diffraction**: light from two coherent point sources produces a stationary interference pattern of alternating, equally spaced and sized bright and dark bands.

Unit 10
Static Electricity and Fields

√

☐ 1. **Static electricity** deals with electrical charges at rest.

☐ 2. In general, matter becomes charged through a transfer of electrons. A rubber rod rubbed with fur acquires a negative charge. A glass rod rubbed with silk acquires a positive charge.

☐ 3. An object that has a deficiency of electrons has a **positive** charge; one with an excess of electrons is **negative**, and one with equal electrons and protons is **neutral**.

☐ 4. Electric charges pass readily through certain substances called **conductors** and pass only with great difficulty through other substances called **insulators**.

☐ 5. **Grounding** either removes or adds electrons to the object to neutralize the charge.

☐ 6. Unlike charges **attract** and like charges **repel**.

☐ 7. The net charge in a closed system is constant (conserved).

☐ 8. An object **charged by contact** acquires the same kind of charge as the charging object. Objects of equal geometry share charges equally.

☐ 9. An object may be **charged by induction.** It acquires a charge opposite to that of the charging object.

☐ 10. A neutral object is attracted by a charged object because of a redistribution of charge (**polarization**) on the neutral object.

☐ 11. The charge of the electron is one negative **elementary charge**. The charge of a proton is one positive elementary charge.

☐ 12. The unit of charge is the **coulomb**.

☐ 13. The force between fixed-point charges is directly proportional to the product of the charges and inversely proportional to the <u>square</u> of the distance between them.

☐ 14. An **electric field** exists around every charged object. The electric field intensity is a vector quantity. The magnitude is equal to the electric force per unit charge at a given point. The direction of the field is the direction of the force on a <u>positive</u> charge. The newton per coulomb, N/C, is the SI unit for electric field strength.

15. ☐ The field around a point charge is radial. The intensity of the field varies inversely with the square of the distance from the point charge.

16. ☐ The electric field around a charged conducting sphere acts as though all the charge were concentrated at the center. The field within a charged conducting sphere is zero.

17. ☐ The field around a uniformly charged rod is radially directed and its intensity varies inversely with the distance from the rod.

18. ☐ The field between two parallel charged plates is essentially uniform if the distance between the plates is small compared to the dimensions of the plates. The force per unit charge is constant between the plates.

19. ☐ The **electric potential** at any point in an electric field is the work per unit charge required to bring one coulomb of positive charge from infinity to that point. Electric potential is a scalar quantity.

20. ☐ The **potential difference** between two points in an electric field is the change in energy per unit charge as a charge is moved from one point to the other.

21. ☐ The **Volt**, V, is the SI unit of electrical potential and potential difference. One volt = 1 joule/coulomb.

22. ☐ The unit of potential difference is the volt. The volt is a potential difference that exists between two points if one joule of work is required to transfer one coulomb of charge from one point to the other against the electric force.

23. ☐ The **electron-volt (eV)** is the energy required to move one elementary charge through a potential difference of one volt.

24. ☐ The intensity of an electric field may be expressed in terms of the change in potential per unit distance.

25. ☐ Millikan measured the forces on charged oil drops in a uniform electric field. He found that the electric forces were always integral multiples of a small constant. Since the force is proportional to the charge, it follows that there is fundamental unit of charge.

26. ☐ The fundamental unit of charge is the charge on an electron or proton.

27. ☐ Be able to solve problems using the following equations:

$$F_e = k\frac{q_1 q_2}{r^2} \qquad E = \frac{F}{q} \qquad V = \frac{W}{q} \qquad E = \frac{V}{d}$$

Unit 11
Current Electricity and Circuits

√

☐ 1. An **electric current** is the flow of electric charge. It is measured by the rate at which electric charge flows past a given point.

☐ 2. The **conductivity** of solids depends on the number of free charges per unit volume, and the mobility of the charges.

☐ 3. **Conductors** are substances in which there are many free electrons.

☐ 4. **Insulators** are substances in which there are few free electrons.

☐ 5. Other conductors include **electrolytes** and **ionized gases**.

☐ 6. A potential difference is required to maintain a flow of charge between two points in a conductor. A **voltmeter** (connected in parallel) is used to measure potential difference.

☐ 7. The SI unit of current is the **Ampere** (A). It is a fundamental unit. A current of one ampere transfers charge at the rate of one coulomb per second. An **ammeter** (connected in series) is used to measure current.

☐ 8. The Joule, J, is the SI unit for electrical energy. It is a derived unit. Electrical energy is a scalar quantity.

☐ 9. **Electric power** is the time rate at which electrical energy is expended.

☐ 10. The watt is the SI unit for electrical power.

☐ 11. **Resistance** is the ratio of the potential difference across a conductor to the current in it. The SI unit of resistance is the ohm (Ω).

☐ 12. The resistance of a conductor of uniform cross-section and composition varies directly as its length and inversely as its cross-sectional area.

☐ 13. In general, the resistance of metals increases with increasing temperature. The resistance of nonmetals and solutions usually decreases with increasing temperature.

☐ 14. A **circuit** is a closed path in which a current can exist.

☐ 15. A **series circuit** is one in which there is only one current path.

☐ 16. A **parallel circuit** is one in which there is more than one current path.

☐ 17. Know the characteristics of series and parallel circuits and be able to draw circuits using the proper circuit symbols.

☐ 18. The algebraic sum of the currents entering any circuit junction is equal to zero (conservation of charge).

✓

☐ **19.** The algebraic sum of all the potential drops and applied voltages around a complete circuit is equal to zero (conservation of energy).

☐ **20.** Be able to solve problems using the following equations:

$$I = \frac{\Delta q}{t} \qquad R = \frac{V}{I} \qquad R = \frac{\rho L}{A}$$

$$P = VI = I^2 R = \frac{V^2}{R} \qquad\qquad W = Pt = VIt = I^2 Rt = \frac{V^2 t}{R}$$

<table>
<tr><td align="center">**Series Circuits**</td><td align="center">**Parallel Circuits**</td></tr>
<tr><td>$I = I_1 = I_2 + I_3 = \dots$</td><td>$I = I_1 + I_2 + I_3 + \dots$</td></tr>
<tr><td>$V = V_1 + V_2 + V_3 + \dots$</td><td>$V = V_1 = V_2 = V_3 = \dots$</td></tr>
<tr><td>$R_{eq} = R_1 + R_2 + R_3 + \dots$</td><td>$\dfrac{I}{R_{eq}} = \dfrac{1}{R_1} + \dfrac{1}{R_2} + \dfrac{1}{R_3} + \dots$</td></tr>
</table>

Unit 12
Electromagnetism

√

☐ 1. All substances exhibit **magnetic** properties. Atoms of magnetic materials are grouped in microscopic clusters called **domains**.

☐ 2. The field around a permanent magnet is due to atomic currents.

    a. The magnetic force is concentrated at regions known as **poles**. Like poles **repel**, unlike poles **attract**.

    b. The lines of **magnetic flux (magnetic field lines)** emerge from the north pole and enter the south pole of magnets.

    c. Magnetic flux lines form closed paths and never cross.

    d. The **flux density** is the number of flux lines per unit area and is proportional to the intensity of the field. In other words the field strength is greatest where the flux lines are closest together.

☐ 3. **Permeability** is the property of a material that changes the flux density in a magnetic field from its value in a vacuum.

    • The more permeable a substance, the more it concentrates a magnetic field.

☐ 4. A **compass** is a small bar magnet that can rotate freely. A compass will align itself with the magnetic field of the earth or another magnet.

☐ 5. The direction of the magnetic field is, by convention, the direction in which the north pole of a **compass** would point in the field.

☐ 6. The earth behaves like a large bar magnet with the south pole of the earth's magnet near the northern end of its axis of rotation, while the north pole of the earth's magnet is near the southern end of its axis.

☐ 7. The earth's magnetic pole in the northern hemisphere is about 1800 km from the north pole of its axis.

☐ 8. The magnetic field created by a current in a straight conductor is shown by concentric circles around the conductor, in a plane perpendicular to the conductor.

☐ 9. The field of a loop or solenoid carrying a current is such that the faces of the loop or solenoid show polarity (creating an electromagnet).

☐ 10. A force is exerted on a current-carrying wire (conductor) in a magnetic field if the conductor is not parallel to the magnetic field. The force is perpendicular to both the field and the current.

Unit 13
Quantum Theory, The Atom, & Nuclear Physics

√

- [ ] 1. Light exhibits the characteristics of waves and particles **(particle-wave duality)**.

- [ ] 2. Interference, polarization, and diffraction can only be explained on the basis of **wave theory**.

- [ ] 3. The **photoelectric effect** is the emission of electrons from an object when certain electromagnetic radiation strikes it.

- [ ] 4. The photoelectric effect can be explained only on the basis of **particle theory**.

- [ ] 5. The rate of emission of photoelectrons depends on the intensity of the incident radiation.

- [ ] 6. The emission of photoelectrons is immediate (no time delay).

- [ ] 7. The maximum kinetic energy of photoelectrons depends only on the frequency of the incident radiation and the nature of the surface.

- [ ] 8. For each photo-emissive material there is a minimum frequency below which no photoelectrons will be emitted. This minimum frequency is called the **threshold frequency**.

- [ ] 9. The energy associated with the threshold frequency is called the work function of the metal (material).

- [ ] 10. The photoelectric effect could be explained by assuming that electromagnetic radiation is **quantized**.

- [ ] 11. A **photon** is a quantum of light energy.

- [ ] 12. The photons of electromagnetic radiation act individually (as particles) and their energies are proportional to their frequencies and, therefore, inversely proportional to their wavelengths.

- [ ] 13. The maximum kinetic energy of released electrons is a linear function of the frequency of the photons. The slope is called **Planck's constant** ($h = 6.62 \times 10^{-34}$ J • s).

- [ ] 14. Both energy and momentum are conserved in photon-particle collisions.

- [ ] 15. The momentum of a photon is inversely proportional to its wavelength.

- [ ] 16. Moving particles have wave properties. The wavelength of a particle is inversely proportional to its momentum.

- [ ] 17. Be able to solve problems using the following equation:
$$E_{photon} = hf = \frac{hc}{\lambda}$$

Unit 14
Models of the Atom

☐ **1.** **Radiation** consists of three types: **Alpha (α)**, **Beta (β)**, and **Gamma (γ)**. Beta radiation consists of high speed electrons, and gamma radiation is energy only – no charge.

☐ **2.** The **alpha particle (α)** is a helium ($^4_2$He), nucleus, which consists of two protons and two neutrons and therefore is positively charged. Alpha particles are emitted by radioactive materials such as radium.

☐ **3.** Most of the alpha particles directed at a thin metal foil pass through without being deflected. Some (approx. 1/5000) are scattered through angles ranging up to $180°$.

☐ **4.** Alpha particles are deflected into hyperbolic paths because of the electrostatic forces between them and the positively charged nuclei of the metal foil.

☐ **5.** On the basis of scattering experiments **(Rutherford Gold Foil Experiment)**, Ernest Rutherford proposed a model in which the positive charge of an atom, and most of the mass, are concentrated in a small (about 1/10000 the atom's overall size) dense core, called the nucleus. Electrons are widely scattered from the **nucleus**, with most of the atom being empty space.

☐ **6.** Rutherford's model did not account for the lack of emission of radiation as electrons move about the nucleus and the unique spectrum of each element.

☐ **7.** The **Bohr model** (Niels Bohr) of the hydrogen atom consists of a positively charged nucleus and a single electron revolving in a circular orbit.

☐ **8.** Bohr's assumptions:
   a. An orbiting electron does not lose energy even though it has a centripetal acceleration.

   b. Only a limited number of specified orbits are permitted. Each orbit represents a particular energy level.

   c. When an electron changes from one energy level to another, a **quantum** of energy equal to the difference between the energies of the two levels is either emitted or absorbed.

☐ **9.** The process of raising the energy of atoms is called **excitation**.

☐ **10.** Excited atoms subsequently release the energy as **photons**.

☐ **11.** The lowest possible energy level is called the **ground state** (stationary state). For hydrogen, this is n = 1. For mercury, this is energy level *a*.

√

☐ **12.** The minimum energy necessary to remove an electron from an atom is called the **ionization energy**. (This is determined by reading the energy of the level the electron is in from the reference tables.)

☐ **13.** Be able to read and interpret the Energy Level Diagrams for Hydrogen and Mercury in the Physics Reference Tables.

☐ **14.** Bohr's model of the atom did not successfully predict other aspects of the hydrogen atom nor can it explain the electron orbits of large atoms having many electrons.

☐ **15.** Each element has a unique **spectrum**.

☐ **16.** **Continuous spectra** are produced by incandescent solids and liquids and by incandescent gases under extremely high pressure.

☐ **17.** Luminous gases and vapors at low pressures produce **line spectra**.

☐ **18.** **Absorption spectra** are produced when atoms absorb those photons whose energies are equal to the energies of photons it can emit when excited.

☐ **19.** There is no specific orbit for an electron as it moves about the nucleus. Instead, there is a region of most probable electron location called a state. The high probability volume for an electron is called an **electron cloud** (Erwin Schröedinger, 1926).

☐ **20.** Be able to solve problems using the following equation:

$$E_{photon} = E_i - E_f = hf$$

Unit 15
Elementary Particles and The Standard Model

1. The particles (protons and neutrons) inside the nucleus are called **nucleons**.

2. The **atomic number** is the number of protons in the nucleus. A neutral atom has an equal number of protons and electrons. The symbol for atomic number is Z. Elements differ from each other in atomic number.

3. The **mass number** is the total number of protons and neutrons (nucleons) in the nucleus. The symbol for mass number is A.

4. The symbols used for atoms and nuclides: $^A_Z Sy$

5. The **strong nuclear force** is the force that holds the nucleons together. It is a strong short-range force more than 100 times stronger than the electromagnetic force.

6. The **universal mass unit (u)** is defined as $\frac{1}{12}$ the mass of an nucleus of carbon 12.
   - $1u = 1.66 \times 10^{-27}$ kg

7. The mass of subatomic particles (in u):
   - electron: 0.0005 u
   - proton: 1.0073 u
   - neutron: 1.0087 u

8. **Mass Defect:** The mass of the nucleus is less than the total mass of its individual nucleons. This difference in mass is equivalent to the energy with which the nucleons are bound.

9. The **binding energy** of the nucleus is the energy that must be supplied to it in order to separate it into its nucleons.

10. The binding energy is the energy equivalent of the mass defect. Einstein's mass-**energy equation ($E=mc^2$)** states that the energy equivalent of mass is proportional to the mass and the speed of light squared.

11. Nucleons that have the same atomic number but different number of neutrons are called **isotopes**.

12. **Radioactivity** is the disintegration of the nuclei of atoms.

13. During the process of radioactive decay mass-energy is conserved.

14. In addition to the proton, neutron, and electron, other subatomic particles are also observed to exist. These subatomic particles include the **neutrino, positron, antiproton, antineutron, meson, boson, lepton,** and **baryon**. It is now known that some nuclear particles are composed of constituent particles called **quarks**.

15. Be able to solve problems using the following equation: $E = mc^2$

Glossary

A

Absorption: Wave energy that is not recovered from a boundary.

Absorption spectrum: A continuous spectrum, like that of white light, interrupted by dark lines or bands. It is produced by the absorption of certain wavelengths by a substance through which the light or other radiation passes.

Acceleration: A vector quantity that represents the time-rate of change in velocity.

Acceleration of gravity: Rate of change of velocity due to gravitational attraction of Earth.

Accepted value: Known value published in reference books.

Accuracy: Closeness of a measurement to the standard value of a quantity.

Action/reaction pair: Force pairs that act together, for example: walking across the floor, your foot pushes on the floor (action) and the floor pushes on your foot (reaction).

Alpha decay: The emission of an alpha particle from a nucleus.

Alpha particles ($_2^4 He$): The nucleus of helium atoms, consisting of 2 protons and 2 neutrons.

Alternating current (AC): A current that reverses its direction at a regular frequency.

Ammeter: A meter for measuring electric current.

Ampere: The SI unit of current. It is a fundamental unit.

Amplitude: The maximum displacement of a vibrating particle or wave from its rest position.

Angle of incidence: The angle between the incident ray and the normal to the surface at the point where the ray strikes the surface.

Angle of reflection: The angle between the reflected ray and the normal to the surface from which it is reflected.

Angle of refraction: The angle between a ray emerging from the surface and the normal to that surface at the point where that ray emerges.

Antielectron ($_{+1}^0 e$): A positive electron or positron.

Antimatter: Matter composed of antiparticles.

Antineutrino: Antiparticle of the neutrino.

Antineutron ($_0^1 n$): Antiparticle of the neutron.

Antinodes: The region or point of maximum amplitude between adjacent nodes of a standing wave.

Antiparticle: The counterpart of sub-nuclear particle of matter, whose main property is that it and the particle annihilate each other on coming together, liberating their energy as radiation.

Antiproton ($_{-1}^1 H$): Antiparticle of the proton, having the same mass but a negative charge.

Applied force: Any force that is acting on an object.

GLOSSARY

Atom: The smallest particle of an element that has all its chemical properties.

Atomic number: is the number of protons in the nucleus. The symbol for the atomic number is Z.

Attractive force: Positive force in magnets and electromagnetic fields.

B

Balmer series: A series of related lines in the visible part of the spectrum of hydrogen. It is produced by the electrons in excited hydrogen atoms which pass from higher energy levels to the one whose quantum number n=2.

Baryons: Nuclear particles with masses equal to or greater than the mass of the proton.

Base units: a set of units for physical quantities from which every other unit can be generated. Also called fundamental units.

Beats: A series of alternate reinforcements and cancellations produced by two sets of superimposed sound waves of close but different frequencies heard as a throbbing effect.

Beta decay: The emission of an electron from a nucleus.

Beta particles: High-speed electrons emitted by a radioactive nucleus.

Binding energy: Energy that must be supplied to a nucleus to separate it into its nucleons. The binding energy is the energy equivalent of the mass defect.

C

Center of mass: the location in which an object will be perfectly balanced horizontally if suspended from that point. It is the location where all of the mass of the system could be considered to be located.

Centripetal acceleration: Acceleration always at right angles to the velocity of a particle, directed toward the center of the circular path.

Centripetal force: Force directed toward the center of a circle; keeps particles moving in uniform circular motion.

Charged particle: A particle with an electric charge like a proton or electron.

Circuit element: Any component that is part of an electric circuit for example a battery, wire or light bulb.

Closed system: An object or collection of objects where there are no forces or energy exchanges from outside the system. Also called an isolated system.

Coefficient of friction: The ratio of the friction force and the formal force.

Coherent light: Light in which all waves leaving the source are in phase.

Compass: A small magnetized pointer free to align itself with a magnetic field of the earth or another magnet.

Complimentary colors: Primary and secondary color which, when added, produce white light.

Components of a vector: Two or more vectors (usually perpendicular) which, when added together, produce the original vector.

Compression: Portion of a longitudinal wave where particles of the medium are squeezed together. Also called a condensation

Concurrent forces: Forces acting on the same point at the same time.

Conductor: A material that has a low resistivity for electric current.

Conduction: The method of transferring electric charge by physically touching one object to another.

Constant force: Any push or pull on an object that does not change in magnitude over time.

Constant velocity: The ratio of distance vs. time does not change. Example a car moving at 55 miles per hour north.

Constructive inference: When two waves meet that are in phase - a crest meets a crest and/or a trough meets a trough - resulting in larger amplitude.

Conversion factor: A fraction, equal to one, that is used to convert a measured quantity to a different unit of measure without changing the relative amount

Coulomb (C): Unit of quantity of electric charge equal to the charge found on 6.25×10^{18} electrons. It is the charge that is transferred by a current of one ampere in one second.

Coulomb's Law: The force between fixed point charges is directly proportional to the product of the charges and inversely proportional to the square of the distance between them.

Crest: The point on a transverse wave with maximum upward vertical displacement from the rest position.

Critical angle: The angle of incidence for which a ray passing obliquely from an optically more dense to an optically less dense medium has an angle of refraction of 90° and hence does not pass through the interface.

Current: A flow of a liquid, or gas, or of particles such as electric charges.

Current electricity: Moving electric charges through a conductor.

Curved mirror: A reflecting surface that is either concave or convex.

Cycle: One complete vibration of any mechanical oscillation or other periodic change.

D

De Broglie principle: Material particles have wavelike characteristics; wavelength varies inversely with momentum

Deceleration: A negative acceleration where the velocity is reduced in the direction of travel.

Dependent variable: The result. It depends on the initial conditions.

Derived unit: Unit of measurement defined in terms of the primary S.I. Units (m/s, m/s^2 , Joule)..

Destructive interference: The superposition of two waves approximately in opposite phase so that their combined amplitude is the difference between their amplitudes and smaller than either.

Diffraction: The spreading of waves around an obstacle and into the region behind it.

Diffraction grating (transmission type): An optically transparent surface on which are ruled thousands of equidistant opaque parallel lines; it uses the diffraction effects of the slits between these lines to separate light passing through it into its spectrum.

Diffuse reflection: The scattering of light rays by a reflection of light from a rough surface.

Direct current (DC): The movement in electrons in one direction around a circuit.

Direct relationship: As one quantity increases, the other quantity also increases in a graphical relationship.

Direction of propagation: The direction that a wave moves through a medium.

Dispersion: The separation of polychromatic light into its component wavelengths.

Dispersive medium: Any material in which light will spread out. For example when white light hits a prism and disperses into the colors of the rainbow.

Displacement: Is a vector quantity that represents the length and direction of a straight line path from one point to another.

Distance: A scalar quantity that represents the length of a path from one point to another.

Domain: A tiny section of a ferromagnetic substance, such as iron, in which the atoms are lined up with their north poles facing one direction. The entire section acts like a single tiny magnet with its own north and south poles.

Doppler effect: The change in frequency or pitch of sound waves, heard when the source of sound and the observer are moving toward or away from each other. A similar change is observed in the frequency or color of light, when the source of light and the observer are moving toward or away from each other.

Dynamics: Deals with the relation between the forces acting on an object and the resulting change in motion.

E

Echo: Rebound of a pulse or wave from a distant surface.

Efficiency: Ratio of output work to input work expressed as a percent.

Elastic collision: A collision in which both the total momentum and the total kinetic energy of the colliding bodies have the same values before and after the collision.

Elastic limit: The largest stress that can be applied to the body without permanently deforming it.

Elastic potential energy: Potential energy stored as a result of deformation of an elastic object, such as the stretching of a spring. It is equal to the work done to stretch the spring.

Elasticity: Ability of an object to return to its original form after removal of deforming forces.

Electric charges: A fundamental property of matter for example electron carry a negative charge, protons carry a positive charge.

Electric circuit: An unbroken path along which electricity flows or is intended to flow.

Electric current: The flow of electric charges, such as electrons in a metallic conductor, or ions in a liquid or gas.

Electric field: A vector field that surrounds a charged body in space. This field represents the physical effect on charged particles in that space.

Electric field intensity: The force exerted by an electric field on a unit charge at that point.

Electrical energy: The ability to do electric work, this is what is paid for on a utility bill.

Electrical power: The rate at which energy is converted to another form such as mechanical energy, heat or light.

Electroweak force: a theory of the unification of two of the four fundamental forces of nature: the electromagnetic and weak forces.

Electromagnet: A coil of wire wound around a soft iron core, whose magnetic field is produced by passing an electric current through the coil.

Electromagnetic energy: Light energy that is partially magnetic and partly electric.

Electromagnetic induction: The process of producing an EMF in a conductor by changing the magnetic flux passing through the circuit.

Electromagnetic radiation: Any of the spectrum of energy released by an accelerating charged particle.

Electromagnetic spectrum: The entire family of electromagnetic radiations ranging from short wavelength, high-energy gamma rays to long-wavelength, low-energy radio waves.

Electromagnetic waves: Transverse waves moving at the speed of light and consisting of rapidly alternating electric and magnetic fields at right angles to each other and to the direction in which the waves are traveling.

Electromagnetism: The study of the forces that occur between electrically charged particles.

Electron: Subatomic particle of small mass and negative charge.

Electron cloud: Region of high probability of finding an electron.

Electron volt: A unit of energy equal to 1.60×10^{-19} joule.

Electron shell: A region surrounding an atomic nucleus which contains orbiting electrons.

Electroscope: Device used to detect the presence of electric charges.

Elementary charge (e): The smallest charge that can be on an object. Equal to 1.6×10^{-19}C.

EMF: The potential difference generated by electromagnetic induction or the voltage produced by a battery or an electric generator.

Emission spectrum: Spectrum produced by the excited atoms of an element.

Energy: Capacity to do work

Equilibrant (E): Force equal in magnitude to a resultant, but opposite in direction.

Equilibrium: Condition in which the net force on an object is zero.

Equivalent resistance: Resistance of a single resistor that could replace a combination of resistors.

Excitation: The addition of a discrete amount of energy (called excitation energy) to an atom that results in electron(s) raising from lowest energy (ground state) to one of higher energy (excited state).

Excited atom: Atom with one or more electrons in a higher than normal energy level

Experimental value: Value measured directly from experiment.

Extrapolation: The continuation of the data trend to predict future results beyond measured points.

F

Ferromagnetism: The ability of iron, nickel, and cobalt to be strongly attracted by magnets.

Field: A region in space where particles are subject to being influenced by forces such as those of gravity, magnetism or electricity.

Field force: A non-contact force acting on a particle at various positions in space.

Field strength: The ratio of force to charge on an object in a magnetic field.

Fluorescence: The process whereby a substance emits radiation (usually as visible light) when struck by charged particles, such as electrons or alpha particles, or when radiation of a higher frequency (usually ultraviolet light) falls on it.

Flux density: The number of magnetic flux lines per unit area. It is the force exerted per unit current per unit length when the current is perpendicular to the field.

Force: A vector quantity that may be defined as a push or pull.

Frame of reference: Coordinate system used to describe motion.

Free-body diagram (FBD): a sketch of an object identifying all of the forces acting on the body shown.

Free fall: An object falling due to a gravitational pull.

Frequency: The number of cycles occurring per unit time.

Friction: A force opposing the relative motion of two objects in contact. When an object moves against friction, work is done.

Fundamental S.I. unit: The seven primary S.I. Units (meter, second, kilogram, ampere, kelvin, mole, candela).

G

Galvanometer: An instrument used to detect and measure very small electric currents.

Gamma rays (γ): Highly penetrating electromagnetic radiations of very short wavelengths emitted by the nucleus of radioactive atoms.

Generator: A device that converts mechanical energy into electrical energy.

Gram: A small unit of mass, equal to 1/1000 of the standard kilogram.

Gravitational field: A model used to explain the influence that an object exerts into the space around itself, producing an attractive force on another object.

Gravitational force: The force of attraction that every mass exerts upon every other mass.

Gravitational mass: The mass determined by measuring an object on an equal arm balance.

Gravitational potential energy: The potential energy an object has due to its elevated position.

Graviton: The particle assumed to be the carrier of the gravitational force.

Ground state: The condition of an atom when its electrons are at the lowest possible energy levels. Also called stationary state.

Grounding: Connecting a charge object to the earth to remove the object's charge.

H

Hadron: A grouping of elementary particles which has an internal structure that contains quarks.

Hertz: Unit of frequency equal to one event (cycle) per second.

Hooke's Law: The strain produced in an elastic body is directly proportional to stress as long as the elastic limit is not exceeded.

Horsepower: A unit of power equal to 746 watts.

Huygens' principal: Every point on a wave front may be considered a source of wavelets with the same speed.

Hypothesis: A plausible but unproved explanation for a scientific observation or phenomenon.

I

Illuminated: an object that does not emit its own light but is visible from ambient light striking it.

Image: The likeness of an object made by a lens or mirror; it may be real or virtual.

Impulse: A vector quantity with a magnitude equal to the product of the unbalanced force and the time the force acts. Its direction is the same as that of the force.

Incident ray: The ray which falls upon a surface between two substances.

Inclined plane: A simple machine consisting essentially of a sloping surface.

Independent variable: The value that the experimenter is free to manipulate.

Index of refraction (n): The ratio of the speed of light in a vacuum (c) to the speed of light in some medium.

Induced EMF: A potential difference applied to a conductor by a changing magnetic field.

Induced magnetism: Magnetism produced in a magnetic substance when brought into the field of a magnet.

Induction: Is a process by which a charged object caused a redistribution of the charges of another object without contact.

Inelastic collision: A collision in which momentum is conserved but kinetic energy is not conserved. In a totally inelastic collision, the two objects collide and stick together.

Inertia: The tendency of an object to maintain its state of motion, objects at rest tend to remain at rest, objects in motion tend to remain in motion. Directly proportional to the mass of the object.

Infrared light: Electromagnetic radiations whose wavelengths lie between those of visible light and radio waves.

Instantaneous velocity: The velocity of an object at any given instant (especially that of an accelerating object).

Insulators: Material through which the flow of electrical charge is greatly inhibited.

Interface: Common boundary between two materials having different properties.

Interference: Is the effect produced by two or move waves which are passing simultaneously through a region.

Internal energy: Is the total kinetic and potential energy associated with the motions and relative positions of the molecules of an object, apart from any kinetic or potential energy of the object as a whole.

Internal reflection: A wave, like light, remains in the medium it entered for example fiber optics.

Interpolate: A method of constructing new data points within the range of a discrete set of known data points.

Inverse relationship: as one quantity increases, the other quantity decreases in a graphical relationship.

Inverse square law: a relationship stating that a specified physical quantity is inversely related to the square of the other physical quantity.

Ion: An atom, or group of atoms, having an unbalanced electric charge.

Ionization energy: The energy required to detach one of its electrons from an atom and thus turn the atom into an ion.

Isolated system: An object or collection of objects where there are no forces or energy exchanges from outside the system. Also called a closed system.

Isotopes: Different forms of an element whose atomic nuclei have the same number of protons but different numbers of neutrons.

J

Joule: The SI unit of work. It is the work done when a force of one newton acts through a distance of one meter. Energy is also measured in joules.

K

Kilogram: The SI unit of mass. It is a fundamental unit.

Kilowatt-hour: Amount of energy equal to 3.6×10^6 joule.

Kinematics: The mathematical methods of describing motion without regard to the forces which produce it.

Kinetic energy: The energy an object has because of its motion. Like all energy, it is a scalar quantity. Kinetic energy is equal to one-half of the product of the mass and the speed squared.

Kirchhoff's laws (rules): Two laws that describe current and voltage relationships in circuits consisting of networks of resistances.

L

Laser: Device for producing intense coherent light.

Law of conservation of charge: Electric charge can be neither created nor destroyed.

Law of conservation of energy: In non-nuclear changes, energy can be neither created nor destroyed.

Law of conservation of momentum: When no resultant external force acts on a system, the total momentum of a system remains unchanged. When bodies interact their total momentum remains unchanged.

Law of universal gravitation: Every particle of matter in the universe attracts every other particle with a force that is directly proportional to the product of their masses and inversely proportional to the square of the distance between them.

Law of work: In an ideal machine, the work output is equal to the work input.

Leptons: Subatomic particles of little or no mass including electrons, muons, and neutrinos.

Light: An electromagnetic disturbance that can produce the sensation of sight.

Light ray: a narrow beam of light that travels in a straight line path.

Line of best fit: Straight line or curve that fairly approximates the relationship between the data points.

Lines of force: Lines drawn to map electric or magnetic fields to show the direction and the intensity of the field from point to point.

Linear motion: Motion, or something moving, in a straight line.

Longitudinal wave: Particles move in the same direction as the propagation of the wave i.e. push a Slinky forward and back.

Luminous: object that emits its own light and is able to be seen.

M

Macroscopic: The scale in which objects are of a size that is measurable and observable.

Magnetic domain: A tiny section of a ferromagnetic substance, such as iron, in which the atoms are lined up with their north poles facing one direction. The entire section acts like a single tiny magnet with its own north and south poles.

Magnetic field: A region where a magnetic force exists around a magnet or any moving charged object.

Magnetic flux density: Number of magnetic flux lines per unit area.

Magnetic flux lines: Imaginary lines indicating the magnitude and direction of a magnetic field.

Magnetic force: Force between two objects due to the magnetic flux of one or both objects.

Magnetic induction: Strength of a magnetic field.

Magnification: The ratio of image size to object size.

Magnitude: Any size or number.

Mass: Quantity of matter in an object measured by its resistance to a change in its motion (inertia).

Mass defect: The amount by which the mass of an atomic nucleus is less than the sum of the masses of its constituent particles

Mass number: Is the total number of protons and neutrons in the nucleus. The symbol for mass number is A.

Matter: Bodies having mass and volume.

Matter waves: de Broglie waves associated with particles, which are responsible for certain wavelike behaviors of those particles.

Maximum frictional force: Usually static friction. The maximum force required to initiate motion.

Mechanical energy: The sum of the kinetic and potential energy.

Mechanical wave: Any wave that needs a medium to travel through i.e. water waves need water, sound waves need air.

Medium: The "stuff" that a wave moves through i.e. water is the medium for a water wave, air is the medium for a sound wave.

Meson: A particle with a mass between that of the electron and that of a proton. It may be positively charged, negatively charged, or neutral.

Meter: The SI unit of length. It is a fundamental unit.

Microscopic: The scale of objects smaller than those that can easily be seen by the naked eye, requiring a lens or microscope to see them clearly.

Molecule: The smallest piece of an element or compound that can exist independently.

Momentum: A vector quantity that represents the product of an objects mass and velocity.

Monochromatic light: Light of a single color or narrow range of electromagnetic frequencies.

Motion diagram: A pictorial description of an object's motion displaying its location at various equally spaced times on the same diagram.

Muon family: A group of four leptons that contains a muon.

N

Natural frequency: The frequency that requires the least amount of energy to continue to vibrate.

Negative acceleration: Usually acceleration that acts to slow down a moving body, or an acceleration in a direction opposite to that of a positive acceleration.

Net force: The sum of all forces acting on an object.

Neutrino: A neutral particle having a mass number of zero.

Neutron ($_0^1$ n): A neutral particle having about the same mass as a proton and present in all atomic nuclei other than ordinary hydrogen.

Newton: The force which will impart to a mass of one kilogram an acceleration of one meter per second per second. It is a derived unit.

Nodal line: Line connecting nodes.

Node: Point in a medium or field that remains unchanged when acted upon by more than one disturbance simultaneously.

Non-S.I. Unit: Centimeter, gram, minute, etc.

Normal: Direction that is perpendicular to a surface.

Nuclear force: A strong force which holds the nucleons together. Nuclear forces operate when the distance between nucleons is less than 10^{-15} meters.

Nucleon: Proton or neutron.

Nucleus: Core of an atom containing the protons and neutrons.

Nuclide: The nucleus of an isotope of any element having a given mass.

O

Obliqe angle: An angle, such as an acute or obtuse angle, that is not a right angle or a multiple of a right angle.

Ohm (Ω): The SI unit of electrical resistance through which it takes 1 Volt to produce a current of 1 ampere.

Ohm's Law: At constant temperature the resistance of a conductor is equal to the ratio of the potential difference applied across it to the current that flow through it.

Opaque: Not transparent or translucent; impenetrable to light; not allowing light to pass through.

Oscillating system: A system that vibrates or moves back and forth. For example the pendulum of a grandfather clock or a weight bouncing up and down on a spring.

Output: The work obtained from a machine equal to the resistance times the distance it moves.

P

Parallel circuit: An electric circuit in which there is more than one current path.

Paramagnetism: The property of a substance that causes it to be weakly attracted by a magnet.

Particle accelerator: An apparatus used to impart very high velocities and energies to charged subatomic particles, such as protons and deuterons, used as projectiles to penetrate atomic nuclei.

GLOSSARY

Particle diagram: A motion diagram eliminating all of the surrounding objects and fine details of a moving object, replacing the object being analyzed with a dot indicating its gross movements.

Pendulum: A mass suspended from a point so that it swings freely by the influence of gravity.

Period: The time required for the completion of a cycle. It is the reciprocal of the frequency.

Periodic wave: A series of regularly repeated disturbances of a field or medium.

Permanent magnet: An object made from a material that is magnetized and creates its own persistent magnetic field.

Permeability: The ability of a substance to concentrate the lines of force when placed in a magnetic field.

Perpendicular components: Vector components acting at right angles to each other.

Phase: The particular point in the cycle of a waveform, measured as an angle in degrees.

Photoelectric effect: The emission of electrons from an object when certain electromagnetic radiation strikes it.

Photoelectron: An electron emitted when a photosensitive material is illuminated by light.

Photon: Is a quantum of light energy.

Photosensitive: Sensitive or responsive to light or other radiant energy in producing the Photoelectric Effect.

Planck's constant: A universal constant (h) relating the energy of a photon to the frequency of the radiation from which it comes. h= 6.63×10^{-34} joule-second.

Plane mirror: A flat reflecting surface.

Polarization: Transverse wave vibration in a single plane. Longitudinal waves cannot be polarized.

Polychromatic light: Light that contains many colors. For example white light that can be broken up in to the colors of a rainbow.

Positron: The antiparticle of the electron. It has the same mass as the electron but a positive instead of a negative charge.

Potential difference: The difference in potential energy per unit charge between two points in an electric field.

Potential energy: The energy an object has because of its position or condition. Under ideal conditions it is equal to the work required to bring the object to that position or condition.

Power: The time-rate of doing work. It is a scalar quantity.

Pressure: The force per unit of area.

Principle of Superposition: Whenever two (or more) waves travel through the same medium at the same time. The waves pass through each other without being disturbed. The net displacement of the medium when the waves pass is the sum of the individual wave displacements

Precision: The degree to which repeated measurements under unchanged conditions show the same results; or the number of digits from which a value is expressed.

Projectile path: Two dimensional curved motion, for example the path a golf ball will take as it is hit off a tee.

Propagate: Any of the ways in which waves travel.

Property: An observable, measurable quality.

Proton: A subatomic particle with the symbol p and a positive electric charge of 1 elementary charge; the proton is a hadron, composed of quarks.

Pulse: A single vibratory disturbance which moves from point to point transferring energy.

Q

Quantization: To limit the possible values of (a magnitude or quantity) to a discrete set of values by quantum mechanical rules.

Quantum: The unit of energy associated with each given frequency of radiation.

Quantum theory: A theory which assumes that radiant energy is emitted and absorbed by matter in discrete minimum packets equal to Planck's constant times the frequency of the radiation.

Quark: Theoretical particles that make up all hadrons.

R

Radioactive dating: A technique used to determine the relative age of materials such as rocks; usually based on a comparison between the observed abundance of a naturally occurring radioactive isotope and its decay products.

Radioactive materials: Materials that exhibit the phenomenon of spontaneous decay of unstable nuclei.

Rarefaction: A decrease in density and pressure in a medium, such as air, caused by the passage of a sound wave. Also called an expansion.

Real image: An image that is formed when rays of light actually intersect at a single point. A real image can be projected on a screen.

Reflection: The turning back of a wave when it hits a surface, i.e. light reflecting or bouncing of a mirror.

Refraction: The bending of a wave as it passes from one medium to another.

Relative motion: The motion of an object as it compares to the motion of other things around it.

Repulsive force: A force that tends to move away from another force.

Resistance: Opposition to flow of electric current.

Resistivity: The ability of a material to create resistance.

Resonance: The frequency of a force applied to an object or system matches the natural frequency of the object or system.

Rest mass: The mass that a body has when it is at rest.

Resultant (R): The single vector whose effect is the same as the combined effects of two or more similar vectors.

S

S.I. System: Standardized units for scientific measurements.

Scalar quantity: A quantity that has magnitude only, no direction is specified.

Schematic diagram: A representation of a system using specific symbols, For example a circuit diagram uses specific symbols to represent wires, batteries, resistors and other circuit elements.

Scientific notation: Expressing numbers in a form: $M \times 10^n$ where $1 \leq M \leq 10$ and *n* is an integer.

Second: The SI unit of time. It is a fundamental unit.

Series circuit: A circuit that provides a single conducting path.

Shadow: The space behind an illuminated object from which it excludes light.

Short circuit: An electric circuit in which the resistance is so low as to permit a dangerously high current.

Significant figures: Digits which reflect the accuracy and precision of physical measurements and are know with certainty.

Simple harmonic motion: A vibratory or periodic motion, in which the force acting on the vibrating body is proportional to its displacement from its central equilibrium position and always acts toward that position.

Slope: The "steepness" of a line on a graph, determined by taking the "rise" divided by the "run" between two points on a line.

Snell's law: A light ray passing from one medium to another is refracted so that the ratio of the sine of the angle of incidence to the sine of the angle of refraction is equal to a constant for all angles of incidence.

Sound waves: Longitudinal waves in air and other material media set up by vibrating bodies.

Speed: A scalar quantity that represents the magnitude of the velocity.

Speed of light: 3.00×10^8 m/s (in vacuum).

Spring constant: The proportionality between the applied force and the compression or elongation of a spring.

Standard Model: A mathematical description of the elementary particles of matter and the electromagnetic, weak, and strong forces by which they interact.

Standing wave: A pattern of wave crests and troughs that remains stationary in a medium when two waves of equal frequency and amplitude pass through the medium in opposite directions.

State: Physical condition of a material.

Static electricity: Electrical charges at rest.

Strong nuclear force: The force of attraction between two nucleons in which the meson serves as the carrier.

Superconductivity: The complete loss of electrical resistance by certain materials when cooled to near absolute zero.

Superposition: The process whereby two or more waves combine their effects when passing through the same parts of a medium at the same time.

Sympathetic vibration: The vibration of a body at its natural frequency set up by the vibration of another body having the same natural frequency.

T

Tangential velocity: The instantaneous linear velocity of a body moving in a circular path; its direction is tangent to the circular path at that point.

Temperature: That property of matter which determines the direction of the exchange of internal energy between objects. The object at lower temperature will gain internal energy. Absolute temperature is directly proportional to the average kinetic energy of random motion of the molecules of an ideal gas.

Theory: An imagined conceptual scheme, mechanism, or model that provides a plausible explanation of a series of experimental observations.

Thermal energy: Internal energy.

Threshold frequency: The minimum frequency of incident light that will eject a photoelectron from a given metal.

Total energy (of a system): The sum of all the energies acting on a system.

Total internal reflection: The total reflection of a beam of light traveling in an optically dense medium when it falls upon the surface of a less dense medium.

Transmitted wave: Energy projected from one place to another.

Transmutation: A change from one isotope to another of the same or different atomic number because of a gain or loss of protons and/or neutrons by the nucleus.

Translucent substance: One through which light is transmitted diffusely so that bodies cannot be seen through it.

Transparent substance: One through which light passes without diffusion so that bodies can be seen through it.

Transverse wave: Waves such as light, in which the vibrations are perpendicular to the direction in which the waves are traveling.

Triboelectric Series: A list of materials indicating the likelihood in which certain materials become electrically charged after they come into contact with another different material through friction.

Tritium ($_1^3$ H): The isotope of hydrogen having a mass number of 3.

Trough: The point on a transverse wave with maximum downward vertical displacement from the rest position.

U

Ultraviolet light: The range of invisible radiations in the electromagnetic spectrum between violet light and X rays.

Unbalanced force: One force is larger than another.

Uniform circular motion: The motion of a body traveling in a circular path at constant speed.

Uniform motion: Displacement at constant velocity.

Universal mass unit: One twelfth the mass of a carbon-12 atom, the symbol is u.

V

Vacuum: A region of empty space.

Valence band: The outermost shell of an atom. When this shell is completely filled with electrons, a material is an insulator.

Valence electron: An electron in an outer incomplete shell of an atom that takes part in ionic and covalent bonding.

Vector addition: The resultant of two vectors can be represented by the diagonal of a parallel diagram constructed with the two vectors as sides.

Vector quantity: A quantity that has magnitude and direction, often shown as graphically as an arrow with definite length and direction.

Velocity: A vector quantity which represents the time-rate of change of displacement.

Virtual image: Likeness of an object which can be seen or photographed but cannot be projected on a screen.

Volt: A potential difference that exists between two points if one joule of work is required to transfer one coulomb of charge from one point to the other in an electric field.

Voltmeter: An instrument used for measuring electrical potential difference (voltage) between two points in an electric circuit.

W

Watt: The SI unit of power. It is equal to one joule per second.

Wave: A series vibratory disturbances that are propagated from a source.

Wave front: The locus of adjacent points of the wave which are in phase.

Wave speed: The velocity at which a wave travels, $v=f\lambda$.

Wave-particle duality: all objects that exhibit both wave and particle properties.

Wavelength: Is the distance between two consecutive points of a wave that are in phase.

Weak force: Force involved in the decay of atomic nuclei and nuclear particles. A type of electromagnetic force.

Weight: Gravitational attraction of Earth or a celestial body for an object. The product of acceleration due to gravity and mass.

Work: A scalar quantity that is equal to the product of the component of force acting in the direction of the motion and the displacement that the object is moved.

Work-energy theorem: A theorem that states that the net work done on an object is equal to the change in total energy of that object.

Work function: The minimum energy needed to eject a photoelectron from a given metal.

W-particle: Carrier of the weak nuclear interaction.

X

X-rays: A range of deeply penetrating electromagnetic radiations whose wavelengths lie between those of ultraviolet light and gamma rays.

Y

y-intercept: The point where the graph of a function intersects with the y-axis of the graph.

NOTES:

THE UNIVERSITY OF THE STATE OF NEW YORK • THE STATE EDUCATION DEPARTMENT • ALBANY, NY 12234

Reference Tables for Physical Setting/PHYSICS
2006 Edition

List of Physical Constants

Name	Symbol	Value
Universal gravitational constant	G	6.67×10^{-11} N•m^2/kg^2
Acceleration due to gravity	g	9.81 m/s^2
Speed of light in a vacuum	c	3.00×10^8 m/s
Speed of sound in air at STP		3.31×10^2 m/s
Mass of Earth		5.98×10^{24} kg
Mass of the Moon		7.35×10^{22} kg
Mean radius of Earth		6.37×10^6 m
Mean radius of the Moon		1.74×10^6 m
Mean distance—Earth to the Moon		3.84×10^8 m
Mean distance—Earth to the Sun		1.50×10^{11} m
Electrostatic constant	k	8.99×10^9 N•m^2/C^2
1 elementary charge	e	1.60×10^{-19} C
1 coulomb (C)		6.25×10^{18} elementary charges
1 electronvolt (eV)		1.60×10^{-19} J
Planck's constant	h	6.63×10^{-34} J•s
1 universal mass unit (u)		9.31×10^2 MeV
Rest mass of the electron	m_e	9.11×10^{-31} kg
Rest mass of the proton	m_p	1.67×10^{-27} kg
Rest mass of the neutron	m_n	1.67×10^{-27} kg

Prefixes for Powers of 10

Prefix	Symbol	Notation
tera	T	10^{12}
giga	G	10^9
mega	M	10^6
kilo	k	10^3
deci	d	10^{-1}
centi	c	10^{-2}
milli	m	10^{-3}
micro	μ	10^{-6}
nano	n	10^{-9}
pico	p	10^{-12}

Approximate Coefficients of Friction

	Kinetic	Static
Rubber on concrete (dry)	0.68	0.90
Rubber on concrete (wet)	0.58	
Rubber on asphalt (dry)	0.67	0.85
Rubber on asphalt (wet)	0.53	
Rubber on ice	0.15	
Waxed ski on snow	0.05	0.14
Wood on wood	0.30	0.42
Steel on steel	0.57	0.74
Copper on steel	0.36	0.53
Teflon on Teflon	0.04	

REFERENCE TABLES

The Electromagnetic Spectrum

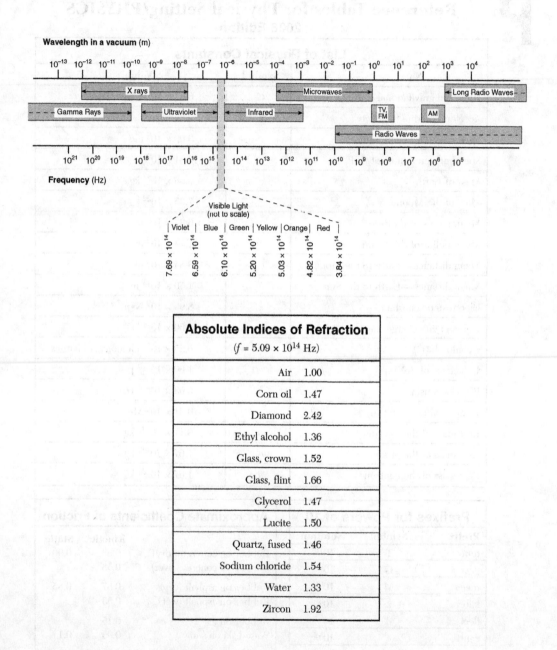

Absolute Indices of Refraction
($f = 5.09 \times 10^{14}$ Hz)

Air	1.00
Corn oil	1.47
Diamond	2.42
Ethyl alcohol	1.36
Glass, crown	1.52
Glass, flint	1.66
Glycerol	1.47
Lucite	1.50
Quartz, fused	1.46
Sodium chloride	1.54
Water	1.33
Zircon	1.92

REFERENCE TABLES

Energy Level Diagrams

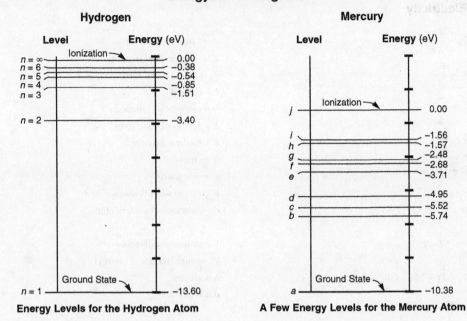

Energy Levels for the Hydrogen Atom

A Few Energy Levels for the Mercury Atom

Classification of Matter

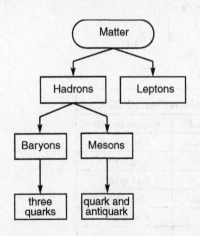

Particles of the Standard Model

Quarks

Name	up	charm	top
Symbol	u	c	t
Charge	$+\frac{2}{3}$e	$+\frac{2}{3}$e	$+\frac{2}{3}$e

	down	strange	bottom
	d	s	b
	$-\frac{1}{3}$e	$-\frac{1}{3}$e	$-\frac{1}{3}$e

Leptons

electron	muon	tau
e	μ	τ
-1e	-1e	-1e

electron neutrino	muon neutrino	tau neutrino
ν_e	ν_μ	ν_τ
0	0	0

Note: For each particle, there is a corresponding antiparticle with a charge opposite that of its associated particle.

Electricity

$$F_e = \frac{kq_1q_2}{r^2}$$

$$E = \frac{F_e}{q}$$

$$V = \frac{W}{q}$$

$$I = \frac{\Delta q}{t}$$

$$R = \frac{V}{I}$$

$$R = \frac{\rho L}{A}$$

$$P = VI = I^2R = \frac{V^2}{R}$$

$$W = Pt = VIt = I^2Rt = \frac{V^2t}{R}$$

A = cross-sectional area
E = electric field strength
F_e = electrostatic force
I = current
k = electrostatic constant
L = length of conductor
P = electrical power
q = charge
R = resistance
R_{eq} = equivalent resistance
r = distance between centers
t = time
V = potential difference
W = work (electrical energy)
Δ = change
ρ = resistivity

Series Circuits

$$I = I_1 = I_2 = I_3 = \ldots$$
$$V = V_1 + V_2 + V_3 + \ldots$$
$$R_{eq} = R_1 + R_2 + R_3 + \ldots$$

Parallel Circuits

$$I = I_1 + I_2 + I_3 + \ldots$$
$$V = V_1 = V_2 = V_3 = \ldots$$
$$\frac{1}{R_{eq}} = \frac{1}{R_1} + \frac{1}{R_2} + \frac{1}{R_3} + \ldots$$

Circuit Symbols

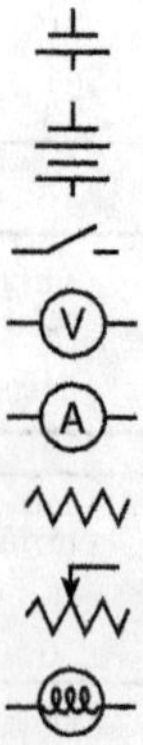

cell

battery

switch

voltmeter

ammeter

resistor

variable resistor

lamp

Resistivities at 20°C

Material	Resistivity ($\Omega \cdot m$)
Aluminum	2.82×10^{-8}
Copper	1.72×10^{-8}
Gold	2.44×10^{-8}
Nichrome	$150. \times 10^{-8}$
Silver	1.59×10^{-8}
Tungsten	5.60×10^{-8}

Waves

$v = f\lambda$

$T = \dfrac{1}{f}$

$\theta_i = \theta_r$

$n = \dfrac{c}{v}$

$n_1 \sin \theta_1 = n_2 \sin \theta_2$

$\dfrac{n_2}{n_1} = \dfrac{v_1}{v_2} = \dfrac{\lambda_1}{\lambda_2}$

c = speed of light in a vacuum
f = frequency
n = absolute index of refraction
T = period
v = velocity or speed
λ = wavelength
θ = angle
θ_i = angle of incidence
θ_r = angle of reflection

Modern Physics

$E_{photon} = hf = \dfrac{hc}{\lambda}$

$E_{photon} = E_i - E_f$

$E = mc^2$

c = speed of light in a vacuum
E = energy
f = frequency
h = Planck's constant
m = mass
λ = wavelength

Geometry and Trigonometry

Rectangle

$A = bh$

Triangle

$A = \frac{1}{2}bh$

Circle

$A = \pi r^2$

$C = 2\pi r$

Right Triangle

$c^2 = a^2 + b^2$

$\sin \theta = \dfrac{a}{c}$

$\cos \theta = \dfrac{b}{c}$

$\tan \theta = \dfrac{a}{b}$

A = area
b = base
C = circumference
h = height
r = radius

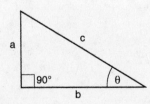

Mechanics

$$\bar{v} = \frac{d}{t}$$

$$a = \frac{\Delta v}{t}$$

$$v_f = v_i + at$$

$$d = v_i t + \frac{1}{2}at^2$$

$$v_f^2 = v_i^2 + 2ad$$

$$A_y = A \sin \theta$$

$$A_x = A \cos \theta$$

$$a = \frac{F_{net}}{m}$$

$$F_f = \mu F_N$$

$$F_g = \frac{Gm_1 m_2}{r^2}$$

$$g = \frac{F_g}{m}$$

$$p = mv$$

$$p_{before} = p_{after}$$

$$J = F_{net}t = \Delta p$$

$$F_s = kx$$

$$PE_s = \frac{1}{2}kx^2$$

$$F_c = ma_c$$

$$a_c = \frac{v^2}{r}$$

$$\Delta PE = mg\Delta h$$

$$KE = \frac{1}{2}mv^2$$

$$W = Fd = \Delta E_T$$

$$E_T = PE + KE + Q$$

$$P = \frac{W}{t} = \frac{Fd}{t} = F\bar{v}$$

a = acceleration
a_c = centripetal acceleration
A = any vector quantity
d = displacement or distance
E_T = total energy
F = force
F_c = centripetal force
F_f = force of friction
F_g = weight or force due to gravity
F_N = normal force
F_{net} = net force
F_s = force on a spring
g = acceleration due to gravity or gravitational field strength
G = universal gravitational constant
h = height
J = impulse
k = spring constant
KE = kinetic energy
m = mass
p = momentum
P = power
PE = potential energy
PE_s = potential energy stored in a spring
Q = internal energy
r = radius or distance between centers
t = time interval
v = velocity or speed
$\bar{v}$ = average velocity or average speed
W = work
x = change in spring length from the equilibrium position
Δ = change
θ = angle
μ = coefficient of friction

Index

NOTES:

Part A
Answer all questions in this part.

Directions (1–35): For *each* statement or question, choose the word or expression that, of those given, best completes the statement or answers the question. Some questions may require the use of the *2006 Edition Reference Tables for Physical Setting/Physics*. Record your answers on your separate answer sheet.

Base your answers to questions 1 and 2 on the information below.

In a drill during basketball practice, a player runs the length of the 30.-meter court and back. The player does this three times in 60. seconds.

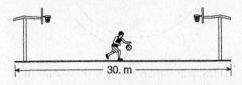

30. m

(Not drawn to scale)

1 The magnitude of the player's total displacement after running the drill is
(1) 0.0 m   (3) 60. m
(2) 30. m   (4) 180 m

2 The average speed of the player during the drill is
(1) 0.0 m/s   (3) 3.0 m/s
(2) 0.50 m/s   (4) 30. m/s

3 A baseball is thrown at an angle of 40.0° above the horizontal. The horizontal component of the baseball's initial velocity is 12.0 meters per second. What is the magnitude of the ball's initial velocity?
(1) 7.71 m/s   (3) 15.7 m/s
(2) 9.20 m/s   (4) 18.7 m/s

4 A particle could have a charge of
(1) 0.8×10^{-19} C   (3) 3.2×10^{-19} C
(2) 1.2×10^{-19} C   (4) 4.1×10^{-19} C

5 Which object has the greatest inertia?
(1) a 15-kg mass traveling at 5.0 m/s
(2) a 10.-kg mass traveling at 10. m/s
(3) a 10.-kg mass traveling at 5.0 m/s
(4) a 5.0-kg mass traveling at 15 m/s

6 A car, initially traveling east with a speed of 5.0 meters per second, is accelerated uniformly at 2.0 meters per second^2 east for 10. seconds along a straight line. During this 10.-second interval the car travels a total distance of
(1) 50. m   (3) 1.0×10^2 m
(2) 60. m   (4) 1.5×10^2 m

7 Which situation describes an object that has *no* unbalanced force acting on it?
(1) an apple in free fall
(2) a satellite orbiting Earth
(3) a hockey puck moving at constant velocity across ice
(4) a laboratory cart moving down a frictionless 30.° incline

8 A child riding a bicycle at 15 meters per second accelerates at −3.0 meters per second^2 for 4.0 seconds. What is the child's speed at the end of this 4.0-second interval?
(1) 12 m/s   (3) 3.0 m/s
(2) 27 m/s   (4) 7.0 m/s

9 An unbalanced force of 40. newtons keeps a 5.0-kilogram object traveling in a circle of radius 2.0 meters. What is the speed of the object?
(1) 8.0 m/s   (3) 16 m/s
(2) 2.0 m/s   (4) 4.0 m/s

10 A 5.00-kilogram block slides along a horizontal, frictionless surface at 10.0 meters per second for 4.00 seconds. The magnitude of the block's momentum is
(1) 200. kg•m/s   (3) 20.0 kg•m/s
(2) 50.0 kg•m/s   (4) 12.5 kg•m/s

11 A 0.50-kilogram puck sliding on a horizontal shuffleboard court is slowed to rest by a frictional force of 1.2 newtons. What is the coefficient of kinetic friction between the puck and the surface of the shuffleboard court?
(1) 0.24   (3) 0.60
(2) 0.42   (4) 4.1

12 A number of 1.0-newton horizontal forces are exerted on a block on a frictionless, horizontal surface. Which top-view diagram shows the forces producing the greatest magnitude of acceleration of the block?

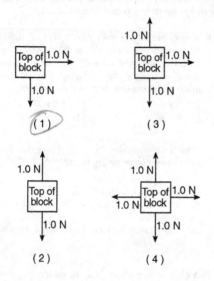

13 On a small planet, an astronaut uses a vertical force of 175 newtons to lift an 87.5-kilogram boulder at constant velocity to a height of 0.350 meter above the planet's surface. What is the magnitude of the gravitational field strength on the surface of the planet?

(1) 0.500 N/kg   (3) 9.81 N/kg
(2) 2.00 N/kg   (4) 61.3 N/kg

14 A car uses its brakes to stop on a level road. During this process, there must be a conversion of kinetic energy into

(1) light energy
(2) nuclear energy
(3) gravitational potential energy
(4) internal energy

15 Which change decreases the resistance of a piece of copper wire?

(1) increasing the wire's length
(2) increasing the wire's resistivity
(3) decreasing the wire's temperature
(4) decreasing the wire's diameter

16 A stone on the end of a string is whirled clockwise at constant speed in a horizontal circle as shown in the diagram below.

Top view

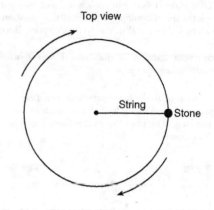

Which pair of arrows best represents the directions of the stone's velocity, v, and acceleration, a, at the position shown?

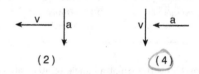

17 How much work is done by the force lifting a 0.1-kilogram hamburger vertically upward at constant velocity 0.3 meter from a table?

(1) 0.03 J   (3) 0.3 J
(2) 0.1 J   (4) 0.4 J

18 Two electrons are separated by a distance of 3.00×10^{-6} meter. What are the magnitude and direction of the electrostatic forces each exerts on the other?

(1) 2.56×10^{-17} N away from each other
(2) 2.56×10^{-17} N toward each other
(3) 7.67×10^{-23} N away from each other
(4) 7.67×10^{-23} N toward each other

19 Which object will have the greatest change in electrical energy?

(1) an electron moved through a potential difference of 2.0 V

(2) a metal sphere with a charge of 1.0×10^{-9} C moved through a potential difference of 2.0 V

(3) an electron moved through a potential difference of 4.0 V

(4) a metal sphere with a charge of 1.0×10^{-9} C moved through a potential difference of 4.0 V

20 The resistance of a circuit remains constant. Which graph best represents the relationship between the current in the circuit and the potential difference provided by the battery?

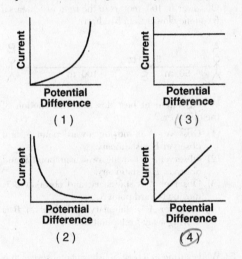

(1)   (3)

(2)   (4)

21 The wavelength of a wave doubles as it travels from medium A into medium B. Compared to the wave in medium A, the wave in medium B has

(1) half the speed

(2) twice the speed

(3) half the frequency

(4) twice the frequency

22 The watt•second is a unit of

(1) power

(2) energy

(3) potential difference

(4) electric field strength

23 Which quantity has both a magnitude and a direction?

(1) energy   (3) power

(2) impulse   (4) work

24 A tuning fork vibrates at a frequency of 512 hertz when struck with a rubber hammer. The sound produced by the tuning fork will travel through the air as a

(1) longitudinal wave with air molecules vibrating parallel to the direction of travel

(2) transverse wave with air molecules vibrating parallel to the direction of travel

(3) longitudinal wave with air molecules vibrating perpendicular to the direction of travel

(4) transverse wave with air molecules vibrating perpendicular to the direction of travel

25 A 3-ohm resistor and a 6-ohm resistor are connected in parallel across a 9-volt battery. Which statement best compares the potential difference across each resistor?

(1) The potential difference across the 6-ohm resistor is the same as the potential difference across the 3-ohm resistor.

(2) The potential difference across the 6-ohm resistor is twice as great as the potential difference across the 3-ohm resistor.

(3) The potential difference across the 6-ohm resistor is half as great as the potential difference across the 3-ohm resistor.

(4) The potential difference across the 6-ohm resistor is four times as great as the potential difference across the 3-ohm resistor.

26 A 3.6-volt battery is used to operate a cell phone for 5.0 minutes. If the cell phone dissipates 0.064 watt of power during its operation, the current that passes through the phone is

(1) 0.018 A   (3) 19 A

(2) 5.3 A   (4) 56 A

27 A monochromatic beam of light has a frequency of 7.69×10^{14} hertz. What is the energy of a photon of this light?

(1) 2.59×10^{-40} J   (3) 5.10×10^{-19} J

(2) 6.92×10^{-31} J   (4) 3.90×10^{-7} J

28 A 3.00×10^{-9}-coulomb test charge is placed near a negatively charged metal sphere. The sphere exerts an electrostatic force of magnitude 6.00×10^{-5} newton on the test charge. What is the magnitude and direction of the electric field strength at this location?

(1) 2.00×10^4 N/C directed away from the sphere
(2) 2.00×10^4 N/C directed toward the sphere
(3) 5.00×10^{-5} N/C directed away from the sphere
(4) 5.00×10^{-5} N/C directed toward the sphere

29 What is characteristic of both sound waves and electromagnetic waves?

(1) They require a medium.
(2) They transfer energy.
(3) They are mechanical waves.
(4) They are longitudinal waves.

30 A small object is dropped through a loop of wire connected to a sensitive ammeter on the edge of a table, as shown in the diagram below.

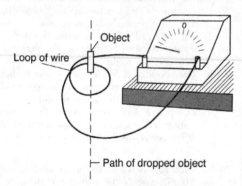

A reading on the ammeter is most likely produced when the object falling through the loop of wire is a

(1) flashlight battery   (3) brass mass
(2) bar magnet   (4) plastic ruler

31 What is the wavelength of a 2.50-kilohertz sound wave traveling at 326 meters per second through air?

(1) 0.130 m   (3) 7.67 m
(2) 1.30 m   (4) 130. m

32 Ultrasound is a medical technique that transmits sound waves through soft tissue in the human body. Ultrasound waves can break kidney stones into tiny fragments, making it easier for them to be excreted without pain. The shattering of kidney stones with specific frequencies of sound waves is an application of which wave phenomenon?

(1) the Doppler effect   (3) refraction
(2) reflection   (4) resonance

33 In the diagram below, a stationary source located at point S produces sound having a constant frequency of 512 hertz. Observer A, 50. meters to the left of S, hears a frequency of 512 hertz. Observer B, 100. meters to the right of S, hears a frequency lower than 512 hertz.

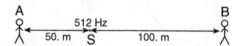

Which statement best describes the motion of the observers?

(1) Observer A is moving toward point S, and observer B is stationary.
(2) Observer A is moving away from point S, and observer B is stationary.
(3) Observer A is stationary, and observer B is moving toward point S.
(4) Observer A is stationary, and observer B is moving away from point S.

34 While sitting in a boat, a fisherman observes that two complete waves pass by his position every 4 seconds. What is the period of these waves?

(1) 0.5 s   (3) 8 s
(2) 2 s   (4) 4 s

35 A wave passes through an opening in a barrier. The amount of diffraction experienced by the wave depends on the size of the opening and the wave's

(1) amplitude   (3) velocity
(2) wavelength   (4) phase

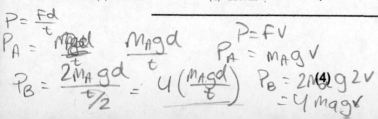

$P = \dfrac{Fd}{t}$

$P_A = \dfrac{M_A gd}{t} \quad \dfrac{M_A gd}{t}$

$P = FV$

$P_A = M_A g V$

$P_B = \dfrac{2 M_A gd}{t/2} = 4\left(\dfrac{M_A gd}{t}\right) \quad P_B = 2M_A g 2V$

$= 4 M_A g V$

Part B–1

Answer all questions in this part.

Directions (36–50): For *each* statement or question, choose the word or expression that, of those given, best completes the statement or answers the question. Some questions may require the use of the *2006 Edition Reference Tables for Physical Setting/Physics*. Record your answers on your separate answer sheet.

36 The length of a football field is closest to

(1) 1000 cm
(2) 1000 dm
(3) 1000 km
(4) 1000 mm

37 A student on an amusement park ride moves in a circular path with a radius of 3.5 meters once every 8.9 seconds. The student moves at an average speed of

(1) 0.39 m/s
(2) 1.2 m/s
(3) 2.5 m/s
(4) 4.3 m/s

38 When a 1.0-kilogram cart moving with a speed of 0.50 meter per second on a horizontal surface collides with a second 1.0-kilogram cart initially at rest, the carts lock together. What is the speed of the combined carts after the collision? [Neglect friction.]

(1) 1.0 m/s
(2) 0.50 m/s
(3) 0.25 m/s
(4) 0 m/s

39 Two elevators, *A* and *B*, move at constant speed. Elevator *B* moves with twice the speed of elevator *A*. Elevator *B* weighs twice as much as elevator *A*. Compared to the power needed to lift elevator *A*, the power needed to lift elevator *B* is

(1) the same
(2) twice as great
(3) half as great
(4) four times as great

40 What is the maximum height to which a motor having a power rating of 20.4 watts can lift a 5.00-kilogram stone vertically in 10.0 seconds?

(1) 0.0416 m
(2) 0.408 m
(3) 4.16 m
(4) 40.8 m

41 What is the current in a wire if 3.4×10^{19} electrons pass by a point in this wire every 60. seconds?

(1) 1.8×10^{-18} A
(2) 3.1×10^{-11} A
(3) 9.1×10^{-2} A
(4) 11 A

42 Which graph represents the relationship between the magnitude of the gravitational force exerted by Earth on a spacecraft and the distance between the center of the spacecraft and center of Earth? [Assume constant mass for the spacecraft.]

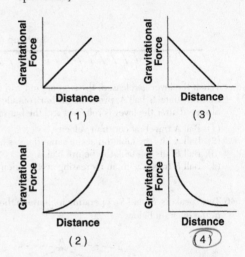

43 To increase the brightness of a desk lamp, a student replaces a 50-watt incandescent lightbulb with a 100-watt incandescent lightbulb. Compared to the 50-watt lightbulb, the 100-watt lightbulb has

(1) less resistance and draws more current
(2) less resistance and draws less current
(3) more resistance and draws more current
(4) more resistance and draws less current

44 Electrons in excited hydrogen atoms are in the $n = 3$ energy level. How many different photon frequencies could be emitted as the atoms return to the ground state?

(1) 1
(2) 2
(3) 3
(4) 4

(5)

45 The diagram below represents a setup for demonstrating motion.

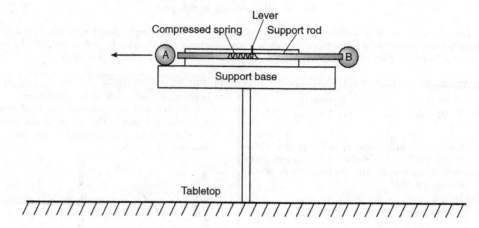

When the lever is released, the support rod withdraws from ball B, allowing it to fall. At the same instant, the rod contacts ball A, propelling it horizontally to the left. Which statement describes the motion that is observed after the lever is released and the balls fall? [Neglect friction.]

(1) Ball A travels at constant velocity.
(2) Ball A hits the tabletop at the same time as ball B.
(3) Ball B hits the tabletop before ball A.
(4) Ball B travels with an increasing acceleration.

46 Two speakers, S_1 and S_2, operating in phase in the same medium produce the circular wave patterns shown in the diagram below.

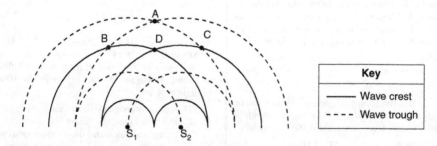

Key	
——	Wave crest
- - - -	Wave trough

At which two points is constructive interference occurring?

(1) A and B
(2) A and D
(3) B and C
(4) B and D

47 A 100.0-kilogram boy and a 50.0-kilogram girl, each holding a spring scale, pull against each other as shown in the diagram below.

The graph below shows the relationship between the magnitude of the force that the boy applies on his spring scale and time.

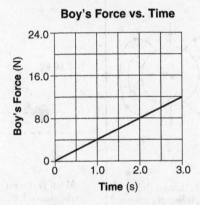

Which graph best represents the relationship between the magnitude of the force that the girl applies on her spring scale and time?

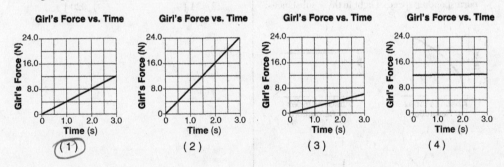

48 In which diagram do the field lines best represent the gravitational field around Earth?

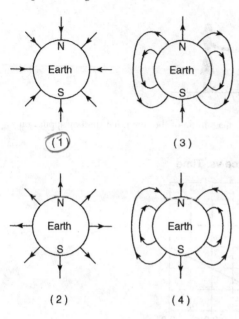

(1)　　　　　　　　　(3)

(2)　　　　　　　　　(4)

49 A ray of light ($f = 5.09 \times 10^{14}$ Hz) travels through various substances. Which graph best represents the relationship between the absolute index of refraction of these substances and the corresponding speed of light in these substances?

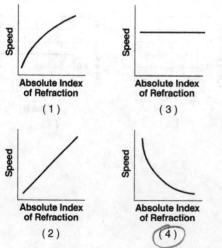

(1)　　　　　　　　　(3)

(2)　　　　　　　　　(4)

50 A pendulum is made from a 7.50-kilogram mass attached to a rope connected to the ceiling of a gymnasium. The mass is pushed to the side until it is at position A, 1.5 meters higher than its equilibrium position. After it is released from rest at position A, the pendulum moves freely back and forth between positions A and B, as shown in the diagram below.

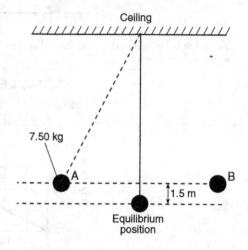

What is the total amount of kinetic energy that the mass has as it swings freely through its equilibrium position? [Neglect friction.]

(1) 11 J　　　　　　(3) 110 J
(2) 94 J　　　　　　(4) 920 J

Part B–2

Answer all questions in this part.

Directions (51–65): Record your answers in the spaces provided in your answer booklet. Some questions may require the use of the *2006 Edition Reference Tables for Physical Setting/Physics.*

Base your answers to questions 51 through 53 on the information below.

A student produced various elongations of a spring by applying a series of forces to the spring. The graph below represents the relationship between the applied force and the elongation of the spring.

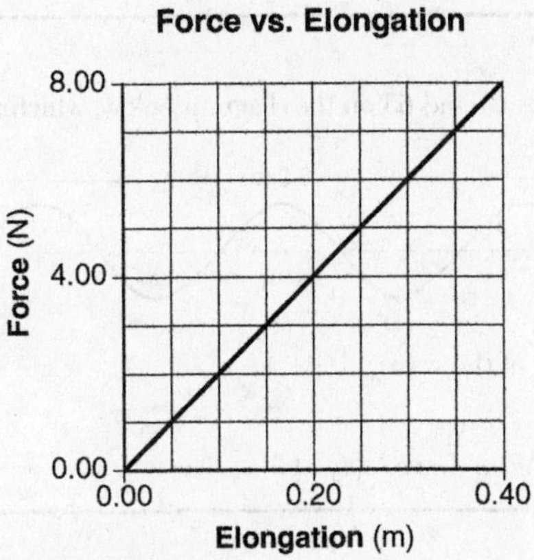

Force vs. Elongation

51 Determine the spring constant of the spring. [1]

$$20 \frac{N}{m}$$

52–53 Calculate the energy stored in the spring when the elongation is 0.30 meter. [Show all work, including the equation and substitution with units.] [2] $PE = \frac{1}{2}kx^2$ $PE = \frac{1}{2}(20)(.3)^2$ $PE = .9J$

54–55 Calculate the time required for a 6000.-newton net force to stop a 1200.-kilogram car initially traveling at 10. meters per second. [Show all work, including the equation and substitution with units.] [2] $\Delta p = Ft$ $0 - (1200)(10) = 6000 t$

$$t = 2 sec$$

56–57 A toy rocket is launched twice into the air from level ground and returns to level ground. The rocket is first launched with initial speed *v* at an angle of 45° above the horizontal. It is launched the second time with the same initial speed, but with the launch angle increased to 60.° above the horizontal. Describe how *both* the total horizontal distance the rocket travels and the time in the air are affected by the increase in launch angle. [Neglect friction.] [2] The total horizontal distance and time are decreased

58–59 Calculate the magnitude of the average gravitational force between Earth and the Moon. [Show all work, including the equation and substitution with units.] [2]

$$F = \frac{Gm_1 m_2}{r^2} \qquad F = \frac{(6.67 \times 10^{-11})(7.35 \times 10^{22})(5.98 \times 10^{24})}{(3.84 \times 10^8)^2}$$

$$F = 1.99 \times 10^{20}$$

Base your answers to questions 60 through 63 on the information below.

A 15-ohm resistor and a 20.-ohm resistor are connected in parallel with a 9.0-volt battery. A single ammeter is connected to measure the total current of the circuit.

60–61 In the space *in your answer booklet*, draw a diagram of this circuit using symbols from the *Reference Tables for Physical Setting/Physics*. [Assume the availability of any number of wires of negligible resistance.] [2]

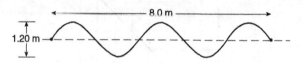

62–63 Calculate the equivalent resistance of the circuit. [Show all work, including the equation and substitution with units.] [2] $\frac{1}{R_T} = \frac{1}{R_1} + \frac{1}{R_2} \cdots$ $\frac{1}{R_T} = \frac{1}{15} + \frac{1}{20}$ $R_T = 8.57$

Base your answers to questions 64 and 65 on the diagram below, which shows a wave in a rope.

```
        |<------------- 8.0 m ------------->|
 1.20 m
```

64 Determine the wavelength of the wave. [1] 3.2m

65 Determine the amplitude of the wave. [1] 0.6m

(10)

Part C

Answer all questions in this part.

Directions (66–85): Record your answers in the spaces provided in your answer booklet. Some questions may require the use of the *2006 Edition Reference Tables for Physical Setting/Physics*.

Base your answers to questions 66 through 70 on the information below.

A runner accelerates uniformly from rest to a speed of 8.00 meters per second. The kinetic energy of the runner was determined at 2.00-meter-per-second intervals and recorded in the data table below.

Data Table

Speed (m/s)	Kinetic Energy (J)
0.00	0.00
2.00	140.
4.00	560.
6.00	1260
8.00	2240

Directions (66–67): Using the information in the data table, construct a graph on the grid *in your answer booklet* following the directions below.

66 Plot the data points for kinetic energy of the runner versus his speed. [1]

67 Draw the line or curve of best fit. [1]

68–69 Calculate the mass of the runner. [Show all work, including the equation and substitution with units.] [2]

70 A soccer player having less mass than the runner also accelerates uniformly from rest to a speed of 8.00 meters per second. Compare the kinetic energy of the less massive soccer player to the kinetic energy of the more massive runner when both are traveling at the same speed. [1]

Base your answers to questions 71 through 75 on the information below.

A river has a current flowing with a velocity of 2.0 meters per second due east. A boat is 75 meters from the north riverbank. It travels at 3.0 meters per second relative to the river and is headed due north. In the diagram below, the vector starting at point P represents the velocity of the boat relative to the river water.

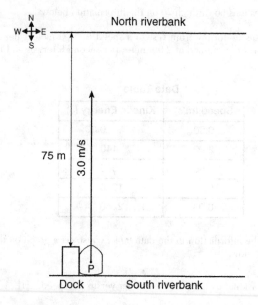

71–72 Calculate the time required for the boat to cross the river. [Show all work, including the equation and substitution with units.] [2]

73 On the diagram *in your answer booklet,* use a ruler and protractor to construct a vector representing the velocity of the river current. Begin the vector at point P and use a scale of 1.0 centimeter = 0.50 meter per second. [1]

74–75 Calculate *or* find graphically the magnitude of the resultant velocity of the boat. [Show all work, including the equation and substitution with units *or* construct the resultant velocity vector *in your answer booklet* for question 73, using a scale of 1.0 centimeter = 0.50 meter per second. The value of the magnitude must be written *in your answer booklet* in the space for questions 74–75.] [2]

Base your answers to questions 76 through 80 on the information below.

A light ray ($f = 5.09 \times 10^{14}$ Hz) is refracted as it travels from water into flint glass. The path of the light ray in the flint glass is shown in the diagram below.

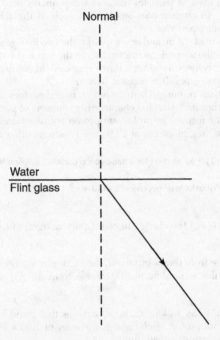

76 Using a protractor, measure the angle of refraction of the light ray in the flint glass. [1]

77–78 Calculate the angle of incidence for the light ray in water. [Show all work, including the equation and substitution with units.] [2]

79 Using a protractor and straightedge, on the diagram *in your answer booklet*, draw the path of the incident light ray in the water. [1]

80 Identify *one* physical event, other than transmission or refraction, that occurs as the light interacts with the water-flint glass boundary. [1]

Base your answers to questions 81 through 85 on the information below.

Two experiments running simultaneously at the Fermi National Accelerator Laboratory in Batavia, Ill., have observed a new particle called the cascade baryon. It is one of the most massive examples yet of a baryon—a class of particles made of three quarks held together by the strong nuclear force—and the first to contain one quark from each of the three known families, or generations, of these elementary particles.

Protons and neutrons are made of up and down quarks, the two first-generation quarks. Strange and charm quarks constitute the second generation, while the top and bottom varieties make up the third. Physicists had long conjectured that a down quark could combine with a strange and a bottom quark to form the three-generation cascade baryon.

On June 13, the scientists running Dzero, one of two detectors at Fermilab's Tevatron accelerator, announced that they had detected characteristic showers of particles from the decay of cascade baryons. The baryons formed in proton-antiproton collisions and lived no more than a trillionth of a second. A week later, physicists at CDF, the Tevatron's other detector, reported their own sighting of the baryon...

Source: D.C., "Pas de deux for a three-scoop particle," *Science News*, Vol. 172, July 7, 2007

81 Which combination of *three* quarks will produce a neutron? [1]

82 What is the magnitude and sign of the charge, in elementary charges, of a cascade baryon? [1]

83 The Tevatron derives its name from teraelectronvolt, the maximum energy it can impart to a particle. Determine the energy, in joules, equivalent to 1.00 teraelectronvolt. [1]

84–85 Calculate the maximum total mass, in kilograms, of particles that could be created in the head-on collision of a proton and an antiproton, each having an energy of 1.60×10^{-7} joule. [Show all work, including the equation and substitution with units.] [2]

The University of the State of New York

REGENTS HIGH SCHOOL EXAMINATION

PHYSICAL SETTING
PHYSICS

————

ANSWER BOOKLET

☐ Male

Student . . . . Sex: ☐ Female

Teacher . . . .

School . . . . Grade . . . .

Record your answers for Part B–2 and Part C in this booklet.

Part B–2
51 _____ N/m
52–53
54–55

56–57 _____

_____

58–59

60–61

62–63

64 _____ m

65 _____ m

Part C

66–67

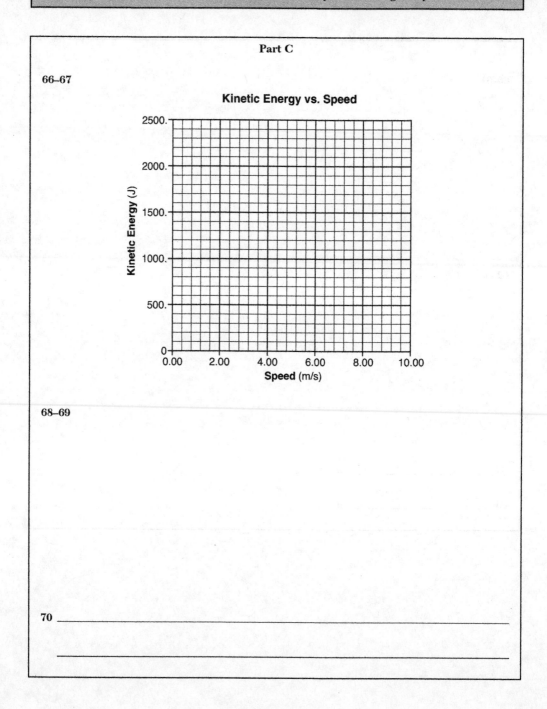

Kinetic Energy vs. Speed

68–69

70 _____

_____

71–72

73

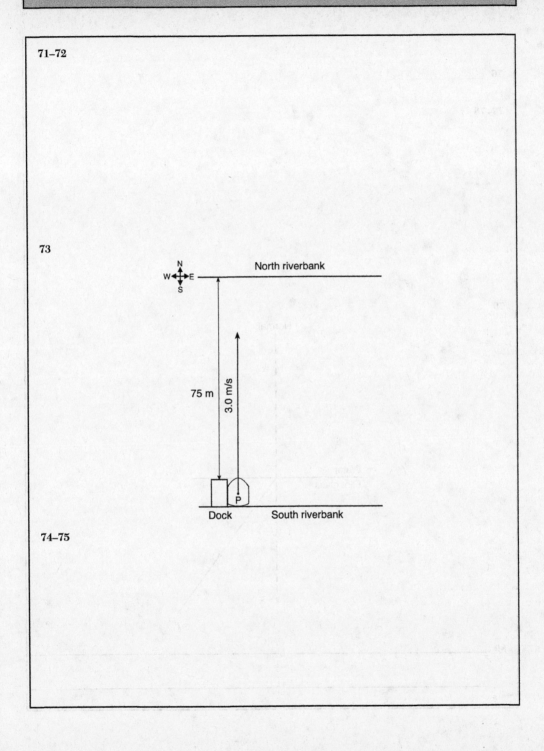

74–75

76 _____ °

77–78

79

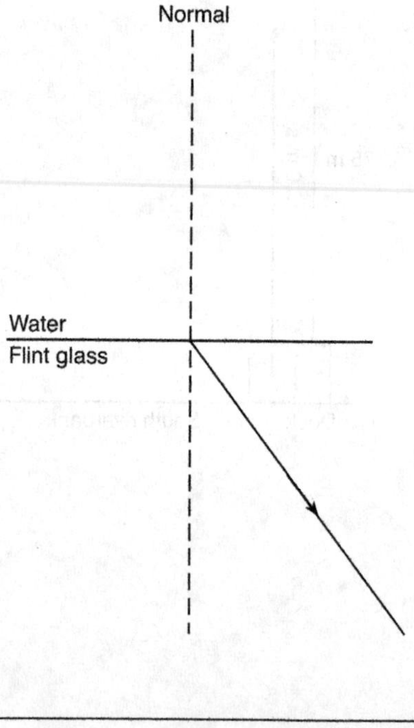

Normal

Water
Flint glass

80 _____

_____

81 _____

82 _____ e

83 _____ J

84–85

Part A

Answer all questions in this part.

Directions (1–35): For *each* statement or question, choose the word or expression that, of those given, best completes the statement or answers the question. Some questions may require the use of the *2006 Edition Reference Tables for Physical Setting/Physics.* Record your answers on your separate answer sheet.

1 Which term identifies a scalar quantity?

(1) displacement   (3) velocity
(2) momentum   (4) time

2 Two 20.-newton forces act concurrently on an object. What angle between these forces will produce a resultant force with the greatest magnitude?

(1) 0°   (3) 90.°
(2) 45°   (4) 180.°

3 A car traveling west in a straight line on a highway decreases its speed from 30.0 meters per second to 23.0 meters per second in 2.00 seconds. The car's average acceleration during this time interval is

(1) 3.5 m/s^2 east   (3) 13 m/s^2 east
(2) 3.5 m/s^2 west   (4) 13 m/s^2 west

4 In a race, a runner traveled 12 meters in 4.0 seconds as she accelerated uniformly from rest. The magnitude of the acceleration of the runner was

(1) 0.25 m/s^2 (3) 3.0 m/s^2
(2) 1.5 m/s^2 (4) 48 m/s^2

5 A projectile is launched at an angle above the ground. The horizontal component of the projectile's velocity, v_x, is initially 40. meters per second. The vertical component of the projectile's velocity, v_y, is initially 30. meters per second. What are the components of the projectile's velocity after 2.0 seconds of flight? [Neglect friction.]

(1) v_x = 40. m/s and v_y = 10. m/s
(2) v_x = 40. m/s and v_y = 30. m/s
(3) v_x = 20. m/s and v_y = 10. m/s
(4) v_x = 20. m/s and v_y = 30. m/s

6 A ball is thrown with an initial speed of 10. meters per second. At what angle above the horizontal should the ball be thrown to reach the greatest height?

(1) 0°   (3) 45°
(2) 30.°   (4) 90.°

7 Which object has the greatest inertia?

(1) a 0.010-kg bullet traveling at 90. m/s
(2) a 30.-kg child traveling at 10. m/s on her bike
(3) a 490-kg elephant walking with a speed of 1.0 m/s
(4) a 1500-kg car at rest in a parking lot

8 An 8.0-newton wooden block slides across a horizontal wooden floor at constant velocity. What is the magnitude of the force of kinetic friction between the block and the floor?

(1) 2.4 N   (3) 8.0 N
(2) 3.4 N   (4) 27 N

9 Which situation represents a person in equilibrium?

(1) a child gaining speed while sliding down a slide
(2) a woman accelerating upward in an elevator
(3) a man standing still on a bathroom scale
(4) a teenager driving around a corner in his car

10 A rock is thrown straight up into the air. At the highest point of the rock's path, the magnitude of the net force acting on the rock is

(1) less than the magnitude of the rock's weight, but greater than zero
(2) greater than the magnitude of the rock's weight
(3) the same as the magnitude of the rock's weight
(4) zero

11 The diagram below shows a compressed spring between two carts initially at rest on a horizontal, frictionless surface. Cart A has a mass of 2 kilograms and cart B has a mass of 1 kilogram. A string holds the carts together.

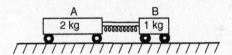

The string is cut and the carts move apart. Compared to the magnitude of the force the spring exerts on cart A, the magnitude of the force the spring exerts on cart B is

(1) the same
(2) half as great
(3) twice as great
(4) four times as great

12 An 8.0-newton block is accelerating down a frictionless ramp inclined at 15° to the horizontal, as shown in the diagram below.

What is the magnitude of the net force causing the block's acceleration?

(1) 0 N
(2) 2.1 N
(3) 7.7 N
(4) 8.0 N

13 At a certain location, a gravitational force with a magnitude of 350 newtons acts on a 70.-kilogram astronaut. What is the magnitude of the gravitational field strength at this location?

(1) 0.20 kg/N
(2) 5.0 N/kg
(3) 9.8 m/s^2
(4) 25 000 N•kg

14 A spring gains 2.34 joules of elastic potential energy as it is compressed 0.250 meter from its equilibrium position. What is the spring constant of this spring?

(1) 9.36 N/m
(2) 18.7 N/m
(3) 37.4 N/m
(4) 74.9 N/m

15 When a teacher shines light on a photocell attached to a fan, the blades of the fan turn. The brighter the light shone on the photocell, the faster the blades turn. Which energy conversion is illustrated by this demonstration?

(1) light → thermal → mechanical
(2) light → nuclear → thermal
(3) light → electrical → mechanical
(4) light → mechanical → chemical

16 Which statement describes a characteristic common to all electromagnetic waves and mechanical waves?

(1) Both types of waves travel at the same speed.
(2) Both types of waves require a material medium for propagation.
(3) Both types of waves propagate in a vacuum.
(4) Both types of waves transfer energy.

17 An electromagnetic wave is produced by charged particles vibrating at a rate of 3.9×10^8 vibrations per second. The electromagnetic wave is classified as

(1) a radio wave
(2) an infrared wave
(3) an x ray
(4) visible light

18 The energy of a sound wave is most closely related to the wave's

(1) frequency
(2) amplitude
(3) wavelength
(4) speed

19 A sound wave traveling eastward through air causes the air molecules to

(1) vibrate east and west
(2) vibrate north and south
(3) move eastward, only
(4) move northward, only

20 What is the speed of light ($f = 5.09 \times 10^{14}$ Hz) in ethyl alcohol?

(1) 4.53×10^{-9} m/s
(2) 2.43×10^2 m/s
(3) 1.24×10^8 m/s
(4) 2.21×10^8 m/s

21 In the diagram below, an ideal pendulum released from position A swings freely to position B.

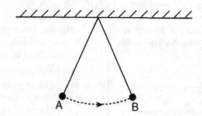

As the pendulum swings from A to B, its total mechanical energy
(1) decreases, then increases
(2) increases, only
(3) increases, then decreases
(4) remains the same

22 The diagram below represents a periodic wave.

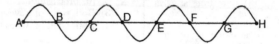

Which two points on the wave are out of phase?
(1) A and C
(2) B and F
(3) C and E
(4) D and G

23 A dry plastic rod is rubbed with wool cloth and then held near a thin stream of water from a faucet. The path of the stream of water is changed, as represented in the diagram below.

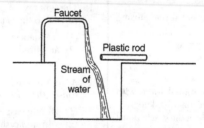

Which force causes the path of the stream of water to change due to the plastic rod?
(1) nuclear
(2) magnetic
(3) electrostatic
(4) gravitational

24 A distance of 1.0×10^{-2} meter separates successive crests of a periodic wave produced in a shallow tank of water. If a crest passes a point in the tank every 4.0×10^{-1} second, what is the speed of this wave?

(1) 2.5×10^{-4} m/s   (3) 2.5×10^{-2} m/s

(2) 4.0×10^{-3} m/s   (4) 4.0×10^{-1} m/s

25 One vibrating 256-hertz tuning fork transfers energy to another 256-hertz tuning fork, causing the second tuning fork to vibrate. This phenomenon is an example of

(1) diffraction   (3) refraction

(2) reflection   (4) resonance

26 Sound waves are produced by the horn of a truck that is approaching a stationary observer. Compared to the sound waves detected by the driver of the truck, the sound waves detected by the observer have a greater

(1) wavelength   (3) period

(2) frequency   (4) speed

27 The electronvolt is a unit of

(1) energy

(2) charge

(3) electric field strength

(4) electric potential difference

28 Which particle would produce a magnetic field?

(1) a neutral particle moving in a straight line

(2) a neutral particle moving in a circle

(3) a stationary charged particle

(4) a moving charged particle

29 A physics student takes her pulse and determines that her heart beats periodically 60 times in 60 seconds. The period of her heartbeat is

(1) 1 Hz   (3) 1 s

(2) 60 Hz   (4) 60 s

30 Moving 4.0 coulombs of charge through a circuit requires 48 joules of electric energy. What is the potential difference across this circuit?

(1) 190 V   (3) 12 V

(2) 48 V   (4) 4.0 V

31 The diagram below shows currents in a segment of an electric circuit.

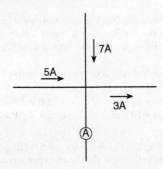

What is the reading of ammeter A?

(1) 1 A   (3) 9 A

(2) 5 A   (4) 15 A

32 An electric dryer consumes 6.0×10^6 joules of electrical energy when operating at 220 volts for 1.8×10^3 seconds. During operation, the dryer draws a current of

(1) 10. A   (3) 9.0×10^2 A

(2) 15 A   (4) 3.3×10^3 A

33 Which net charge could be found on an object?

(1) $+4.80 \times 10^{-19}$ C   (3) -2.40×10^{-19} C

(2) $+2.40 \times 10^{-19}$ C   (4) -5.60×10^{-19} C

34 A photon is emitted as the electron in a hydrogen atom drops from the $n = 5$ energy level directly to the $n = 3$ energy level. What is the energy of the emitted photon?

(1) 0.85 eV   (3) 1.51 eV

(2) 0.97 eV   (4) 2.05 eV

35 In a process called pair production, an energetic gamma ray is converted into an electron and a positron. It is *not* possible for a gamma ray to be converted into two electrons because

(1) charge must be conserved

(2) momentum must be conserved

(3) mass-energy must be conserved

(4) baryon number must be conserved

Part B–1

Answer all questions in this part.

Directions (36–50): For *each* statement or question, choose the word or expression that, of those given, best completes the statement or answers the question. Some questions may require the use of the *2006 Edition Reference Tables for Physical Setting/Physics*. Record your answers on your separate answer sheet.

36 The approximate length of an unsharpened No. 2 pencil is

(1) 2.0×10^{-2} m   (3) 2.0×10^{0} m
(2) 2.0×10^{-1} m   (4) 2.0×10^{1} m

37 The diagram below shows an 8.0-kilogram cart moving to the right at 4.0 meters per second about to make a head-on collision with a 4.0-kilogram cart moving to the left at 6.0 meters per second.

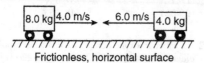

Frictionless, horizontal surface

After the collision, the 4.0-kilogram cart moves to the right at 3.0 meters per second. What is the velocity of the 8.0-kilogram cart after the collision?

(1) 0.50 m/s left   (3) 5.5 m/s left
(2) 0.50 m/s right   (4) 5.5 m/s right

38 Four forces act concurrently on a block on a horizontal surface as shown in the diagram below.

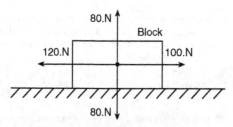

As a result of these forces, the block

(1) moves at constant speed to the right
(2) moves at constant speed to the left
(3) accelerates to the right
(4) accelerates to the left

39 If a motor lifts a 400.-kilogram mass a vertical distance of 10. meters in 8.0 seconds, the *minimum* power generated by the motor is

(1) 3.2×10^{2} W   (3) 4.9×10^{3} W
(2) 5.0×10^{2} W   (4) 3.2×10^{4} W

40 A 4.0-kilogram object is accelerated at 3.0 meters per second² north by an unbalanced force. The same unbalanced force acting on a 2.0-kilogram object will accelerate this object toward the north at

(1) 12 m/s²   (3) 3.0 m/s²
(2) 6.0 m/s²   (4) 1.5 m/s²

41 An electron is located in an electric field of magnitude 600. newtons per coulomb. What is the magnitude of the electrostatic force acting on the electron?

(1) 3.75×10^{21} N   (3) 9.60×10^{-17} N
(2) 6.00×10^{2} N   (4) 2.67×10^{-22} N

42 The current in a wire is 4.0 amperes. The time required for 2.5×10^{19} electrons to pass a certain point in the wire is

(1) 1.0 s   (3) 0.50 s
(2) 0.25 s   (4) 4.0 s

43 When two point charges of magnitude q_1 and q_2 are separated by a distance, r, the magnitude of the electrostatic force between them is F. What would be the magnitude of the electrostatic force between point charges $2q_1$ and $4q_2$ when separated by a distance of $2r$?

(1) F (3) $16F$
(2) $2F$ (4) $4F$

44 The composition of a meson with a charge of −1 elementary charge could be

(1) $s\bar{c}$ (3) $u\bar{b}$
(2) $d\,s\,s$ (4) $\bar{u}\bar{c}\,\bar{d}$

45 Which graph represents the relationship between the kinetic energy and the speed of a freely falling object?

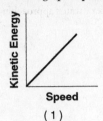 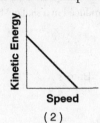  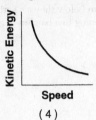

(1)　　　　　(2)　　　　　(3)　　　　　(4)

46 Which diagram represents the electric field between two oppositely charged conducting spheres?

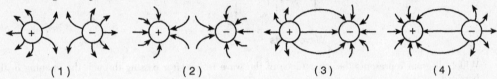

(1)　　　　　(2)　　　　　(3)　　　　　(4)

47 Which graph represents the relationship between the magnitude of the gravitational force, F_g, between two masses and the distance, r, between the centers of the masses?

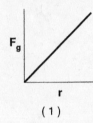 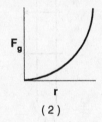 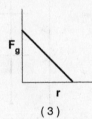 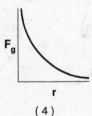

(1)　　　　　(2)　　　　　(3)　　　　　(4)

48 The diagram below shows two waves traveling toward each other at equal speed in a uniform medium.

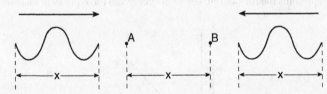

When both waves are in the region between points A and B, they will undergo

(1) diffraction　　　　　　　　　　(3) destructive interference
(2) the Doppler effect　　　　　　　(4) constructive interference

49 The diagram below shows a series of straight wave fronts produced in a shallow tank of water approaching a small opening in a barrier.

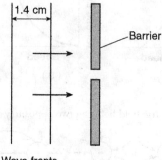

Wave fronts

Which diagram represents the appearance of the wave fronts after passing through the opening in the barrier?

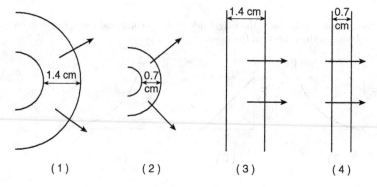

(1)   (2)   (3)   (4)

50 The graph below represents the relationship between energy and the equivalent mass from which it can be converted.

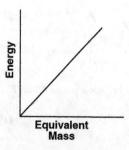

The slope of this graph represents

(1) c

(2) c^2

(3) g

(4) g^2

Part B–2

Answer all questions in this part.

Directions (51–65): Record your answers in the spaces provided in your answer booklet. Some questions may require the use of the *2006 Edition Reference Tables for Physical Setting/Physics*.

51–52 A 25.0-meter length of platinum wire with a cross-sectional area of 3.50×10^{-6} meter^2 has a resistance of 0.757 ohm at 20°C. Calculate the resistivity of the wire. [Show all work, including the equation and substitution with units.]   [2]

53 The diagram below represents a periodic wave moving along a rope.

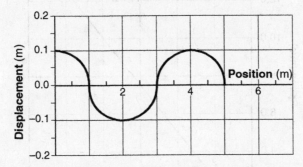

On the grid *in your answer booklet*, draw *at least one* full wave with the same amplitude and half the wavelength of the given wave.   [1]

54–55 A baseball bat exerts an average force of 600. newtons east on a ball, imparting an impulse of 3.6 newton•seconds east to the ball. Calculate the amount of time the baseball bat is in contact with the ball. [Show all work, including the equation and substitution with units.]   [2]

56 The diagram below shows the north pole of one bar magnet located near the south pole of another bar magnet.

N   S

On the diagram *in your answer booklet*, draw *three* magnetic field lines in the region between the magnets.   [1]

Base your answers to questions 57 through 59 on the information and graph below.

The graph below shows the relationship between speed and elapsed time for a car moving in a straight line.

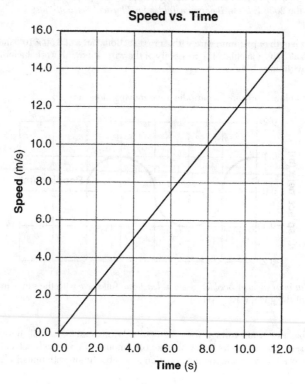

Speed vs. Time

57 Determine the magnitude of the acceleration of the car. [1]

58–59 Calculate the total distance the car traveled during the time interval 4.0 seconds to 8.0 seconds. [Show all work, including the equation and substitution with units.] [2]

Base your answers to questions 60 through 62 on the information below.

A 20.-ohm resistor, R_1, and a resistor of unknown resistance, R_2, are connected in parallel to a 30.-volt source, as shown in the circuit diagram below. An ammeter in the circuit reads 2.0 amperes.

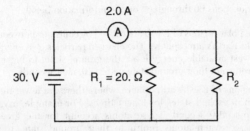

60 Determine the equivalent resistance of the circuit. [1]

61–62 Calculate the resistance of resistor R_2. [Show all work, including the equation and substitution with units.] [2]

Base your answers to questions 63 through 65 on the information below.

A 28-gram rubber stopper is attached to a string and whirled clockwise in a horizontal circle with a radius of 0.80 meter. The diagram in your answer booklet represents the motion of the rubber stopper. The stopper maintains a constant speed of 2.5 meters per second.

63–64 Calculate the magnitude of the centripetal acceleration of the stopper. [Show all work, including the equation and substitution with units.] [2]

65 On the diagram *in your answer booklet*, draw an arrow showing the direction of the centripetal force acting on the stopper when it is at the position shown. [1]

Part C

Answer all questions in this part.

Directions (66–85): Record your answers in the spaces provided in your answer booklet. Some questions may require the use of the *2006 Edition Reference Tables for Physical Setting/Physics*.

Base your answers to questions 66 through 69 on the information below.

Auroras over the polar regions of Earth are caused by collisions between charged particles from the Sun and atoms in Earth's atmosphere. The charged particles give energy to the atoms, exciting them from their lowest available energy level, the ground state, to higher energy levels, excited states. Most atoms return to their ground state within 10. nanoseconds.

In the higher regions of Earth's atmosphere, where there are fewer interatom collisions, a few of the atoms remain in excited states for longer times. For example, oxygen atoms remain in an excited state for up to 1.0 second. These atoms account for the greenish and red glows of the auroras. As these oxygen atoms return to their ground state, they emit green photons ($f = 5.38 \times 10^{14}$ Hz) and red photons ($f = 4.76 \times 10^{14}$ Hz). These emissions last long enough to produce the changing aurora phenomenon.

66 What is the order of magnitude of the time, in seconds, that most atoms spend in an excited state? [1]

67–68 Calculate the energy of a photon, in joules, that accounts for the red glow of the aurora. [Show all work, including the equation and substitution with units.] [2]

69 Explain what is meant by an atom being in its ground state. [1]

Base your answers to questions 70 through 75 on the information below.

A girl rides her bicycle 1.40 kilometers west, 0.70 kilometer south, and 0.30 kilometer east in 12 minutes. The vector diagram in your answer booklet represents the girl's first two displacements in sequence from point P. The scale used in the diagram is 1.0 centimeter = 0.20 kilometer.

70–71 On the vector diagram *in your answer booklet*, using a ruler and a protractor, construct the following vectors:

- Starting at the arrowhead of the second displacement vector, draw a vector to represent the 0.30 kilometer east displacement. Label the vector with its magnitude. [1]

- Draw the vector representing the resultant displacement of the girl for the entire bicycle trip *and* label the vector R. [1]

72–73 Calculate the girl's average speed for the entire bicycle trip. [Show all work, including the equation and substitution with units.] [2]

74 Determine the magnitude of the girl's resultant displacement for the entire bicycle trip, in kilometers. [1]

75 Determine the measure of the angle, in degrees, between the resultant and the 1.40-kilometer displacement vector. [1]

_____

Base your answers to questions 76 through 80 on the information below.

A light ray with a frequency of 5.09×10^{14} hertz traveling in water has an angle of incidence of 35° on a water-air interface. At the interface, part of the ray is reflected from the interface and part of the ray is refracted as it enters the air.

76 What is the angle of reflection of the light ray at the interface? [1]

77 On the diagram *in your answer booklet*, using a protractor and a straightedge, draw the reflected ray. [1]

78–79 Calculate the angle of refraction of the light ray as it enters the air. [Show all work, including the equation and substitution with units.] [2]

80 Identify *one* characteristic of this light ray that is the same in *both* the water and the air. [1]

_____

Base your answers to questions 81 through 85 on the information and diagram below.

A 30.4-newton force is used to slide a 40.0-newton crate a distance of 6.00 meters at constant speed along an incline to a vertical height of 3.00 meters.

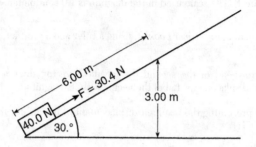

81 Determine the total work done by the 30.4-newton force in sliding the crate along the incline.   [1]

82–83 Calculate the total increase in the gravitational potential energy of the crate after it has slid 6.00 meters along the incline. [Show all work, including the equation and substitution with units.]   [2]

84 State what happens to the kinetic energy of the crate as it slides along the incline.   [1]

85 State what happens to the internal energy of the crate as it slides along the incline.   [1]

The University of the State of New York

REGENTS HIGH SCHOOL EXAMINATION

PHYSICAL SETTING
PHYSICS

ANSWER BOOKLET

☐ Male

Student . . . .   Sex: ☐ Female

Teacher . . . .

School . . . .   Grade . . . .

Record your answers for Part B–2 and Part C in this booklet.

Part B–2

51–52

53

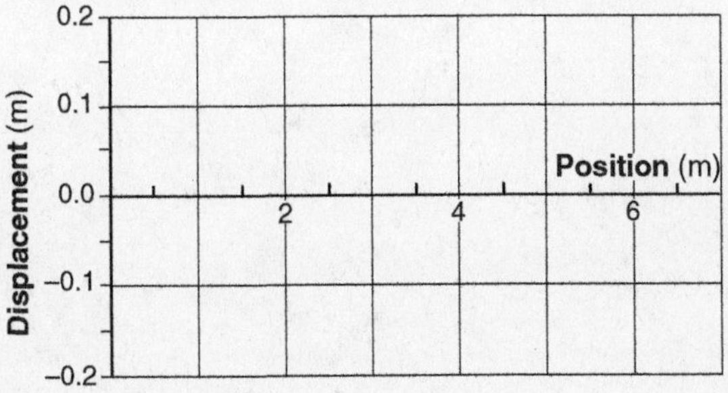

54–55

56

| N | | S |

57 _____ m/s^2

58–59

60 _____ Ω

61–62

63–64

65

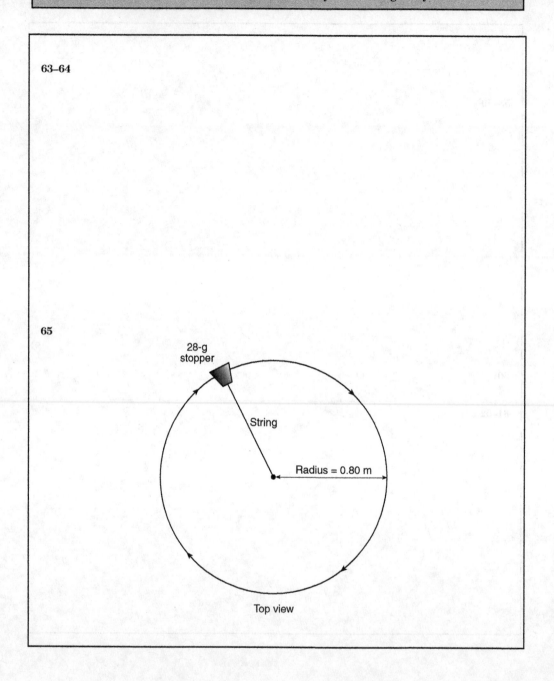

28-g
stopper

String

Radius = 0.80 m

Top view

Part C

66 _____

67–68

69 _____

_____

70–71

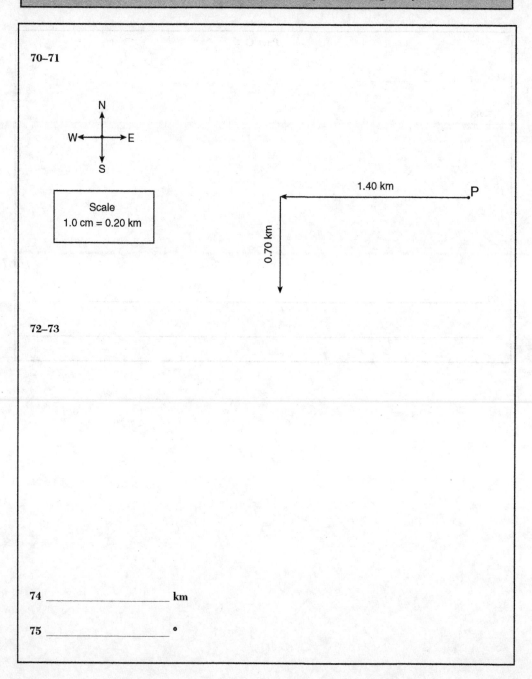

N

W ← → E

S

Scale
1.0 cm = 0.20 km

1.40 km

0.70 km

P

72–73

74 _____ km

75 _____ °

76 _____ °

77

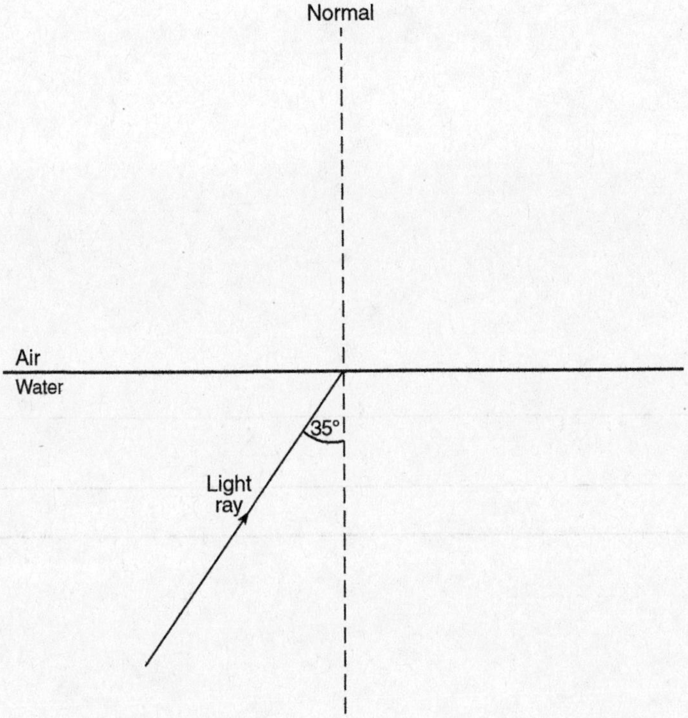

Normal

Air
Water

35°

Light
ray

78–79

80 _____

81 _____ **J**

82–83

84 _____

85 _____

Part A

Answer all questions in this part.

Directions (1–35): For *each* statement or question, choose the word or expression that, of those given, best completes the statement or answers the question. Some questions may require the use of the *2006 Edition Reference Tables for Physical Setting/Physics*. Record your answers on your separate answer sheet.

1 Which quantity is scalar?
(1) mass
(2) force
(3) momentum
(4) acceleration

2 What is the final speed of an object that starts from rest and accelerates uniformly at 4.0 meters per second^2 over a distance of 8.0 meters?
(1) 8.0 m/s
(2) 16 m/s
(3) 32 m/s
(4) 64 m/s

3 The components of a 15-meters-per-second velocity at an angle of 60.° above the horizontal are
(1) 7.5 m/s vertical and 13 m/s horizontal
(2) 13 m/s vertical and 7.5 m/s horizontal
(3) 6.0 m/s vertical and 9.0 m/s horizontal
(4) 9.0 m/s vertical and 6.0 m/s horizontal

4 What is the time required for an object starting from rest to fall freely 500. meters near Earth's surface?
(1) 51.0 s
(2) 25.5 s
(3) 10.1 s
(4) 7.14 s

5 A baseball bat exerts a force of magnitude F on a ball. If the mass of the bat is three times the mass of the ball, the magnitude of the force of the ball on the bat is
(1) F
(2) $2F$
(3) $3F$
(4) $F/3$

6 A 2.0-kilogram mass is located 3.0 meters above the surface of Earth. What is the magnitude of Earth's gravitational field strength at this location?
(1) 4.9 N/kg
(2) 2.0 N/kg
(3) 9.8 N/kg
(4) 20. N/kg

7 A truck, initially traveling at a speed of 22 meters per second, increases speed at a constant rate of 2.4 meters per second^2 for 3.2 seconds. What is the total distance traveled by the truck during this 3.2-second time interval?
(1) 12 m
(2) 58 m
(3) 70. m
(4) 83 m

8 A 750-newton person stands in an elevator that is accelerating downward. The upward force of the elevator floor on the person must be
(1) equal to 0 N
(2) less than 750 N
(3) equal to 750 N
(4) greater than 750 N

9 A 3.0-kilogram object is acted upon by an impulse having a magnitude of 15 newton•seconds. What is the magnitude of the object's change in momentum due to this impulse?
(1) 5.0 kg•m/s
(2) 15 kg•m/s
(3) 3.0 kg•m/s
(4) 45 kg•m/s

10 An air bag is used to safely decrease the momentum of a driver in a car accident. The air bag reduces the magnitude of the force acting on the driver by
(1) increasing the length of time the force acts on the driver
(2) decreasing the distance over which the force acts on the driver
(3) increasing the rate of acceleration of the driver
(4) decreasing the mass of the driver

11 An electron moving at constant speed produces
(1) a magnetic field, only
(2) an electric field, only
(3) both a magnetic and an electric field
(4) neither a magnetic nor an electric field

12 A beam of electrons passes through an electric field where the magnitude of the electric field strength is 3.00×10^3 newtons per coulomb. What is the magnitude of the electrostatic force exerted by the electric field on each electron in the beam?

(1) 5.33×10^{-23} N   (3) 3.00×10^3 N
(2) 4.80×10^{-16} N   (4) 1.88×10^{22} N

13 How much work is required to move 3.0 coulombs of electric charge a distance of 0.010 meter through a potential difference of 9.0 volts?

(1) 2.7×10^3 J   (3) 3.0 J
(2) 27 J   (4) 3.0×10^{-2} J

14 What is the resistance of a 20.0-meter-long tungsten rod with a cross-sectional area of 1.00×10^{-4} meter^2 at 20°C?

(1) $2.80 \times 10^{-5}\ \Omega$ (3) $89.3\ \Omega$
(2) $1.12 \times 10^{-2}\ \Omega$ (4) $112\ \Omega$

15 Two pieces of flint rock produce a visible spark when they are struck together. During this process, mechanical energy is converted into

(1) nuclear energy and electromagnetic energy
(2) internal energy and nuclear energy
(3) electromagnetic energy and internal energy
(4) elastic potential energy and nuclear energy

16 A 15-kilogram cart is at rest on a horizontal surface. A 5-kilogram box is placed in the cart. Compared to the mass and inertia of the cart, the cart-box system has

(1) more mass and more inertia
(2) more mass and the same inertia
(3) the same mass and more inertia
(4) less mass and more inertia

17 Transverse waves are to radio waves as longitudinal waves are to

(1) light waves   (3) ultraviolet waves
(2) microwaves   (4) sound waves

18 As a monochromatic light ray passes from air into water, two characteristics of the ray that will *not* change are

(1) wavelength and period
(2) frequency and period
(3) wavelength and speed
(4) frequency and speed

19 When a mass is placed on a spring with a spring constant of 60.0 newtons per meter, the spring is compressed 0.500 meter. How much energy is stored in the spring?

(1) 60.0 J   (3) 15.0 J
(2) 30.0 J   (4) 7.50 J

20 A boy pushes his sister on a swing. What is the frequency of oscillation of his sister on the swing if the boy counts 90. complete swings in 300. seconds?

(1) 0.30 Hz   (3) 1.5 Hz
(2) 2.0 Hz   (4) 18 Hz

21 What is the period of a sound wave having a frequency of 340. hertz?

(1) 3.40×10^2 s   (3) 9.73×10^{-1} s
(2) 1.02×10^0 s   (4) 2.94×10^{-3} s

22 An MP3 player draws a current of 0.120 ampere from a 3.00-volt battery. What is the total charge that passes through the player in 900. seconds?

(1) 324 C   (3) 5.40 C
(2) 108 C   (4) 1.80 C

23 A beam of light has a wavelength of 4.5×10^{-7} meter in a vacuum. The frequency of this light is

(1) 1.5×10^{-15} Hz   (3) 1.4×10^2 Hz
(2) 4.5×10^{-7} Hz   (4) 6.7×10^{14} Hz

24 When x-ray radiation and infrared radiation are traveling in a vacuum, they have the same

(1) speed   (3) wavelength
(2) frequency   (4) energy per photon

25 The diagram below represents two identical pulses approaching each other in a uniform medium.

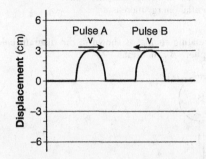

As the pulses meet and are superposed, the maximum displacement of the medium is

(1) –6 cm   (3) 3 cm
(2) 0 cm   (4) 6 cm

26 As a car approaches a pedestrian crossing the road, the driver blows the horn. Compared to the sound wave emitted by the horn, the sound wave detected by the pedestrian has a

(1) higher frequency and a lower pitch
(2) higher frequency and a higher pitch
(3) lower frequency and a higher pitch
(4) lower frequency and a lower pitch

27 When air is blown across the top of an open water bottle, air molecules in the bottle vibrate at a particular frequency and sound is produced. This phenomenon is called

(1) diffraction   (3) resonance
(2) refraction   (4) the Doppler effect

28 An antibaryon composed of two antiup quarks and one antidown quark would have a charge of

(1) +1e   (3) 0e
(2) –1e   (4) –3e

29 Which force is responsible for producing a stable nucleus by opposing the electrostatic force of repulsion between protons?

(1) strong   (3) frictional
(2) weak   (4) gravitational

30 What is the total energy released when 9.11×10^{-31} kilogram of mass is converted into energy?

(1) 2.73×10^{-22} J   (3) 9.11×10^{-31} J
(2) 8.20×10^{-14} J   (4) 1.01×10^{-47} J

31 A shopping cart slows as it moves along a level floor. Which statement describes the energies of the cart?

(1) The kinetic energy increases and the gravitational potential energy remains the same.
(2) The kinetic energy increases and the gravitational potential energy decreases.
(3) The kinetic energy decreases and the gravitational potential energy remains the same.
(4) The kinetic energy decreases and the gravitational potential energy increases.

32 Two identically-sized metal spheres, A and B, are on insulating stands, as shown in the diagram below. Sphere A possesses an excess of 6.3×10^{10} electrons and sphere B is neutral.

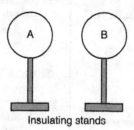

Which diagram best represents the charge distribution on sphere B?

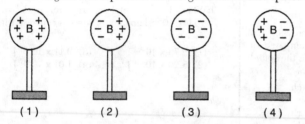

(1)   (2)   (3)   (4)

33 Two points, A and B, are located within the electric field produced by a –3.0 nanocoulomb charge. Point A is 0.10 meter to the left of the charge and point B is 0.20 meter to the right of the charge, as shown in the diagram below.

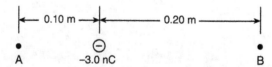

Compared to the magnitude of the electric field strength at point A, the magnitude of the electric field strength at point B is

(1) half as great
(2) twice as great

(3) one-fourth as great
(4) four times as great

34 The diagram below represents two waves, A and B, traveling through the same uniform medium.

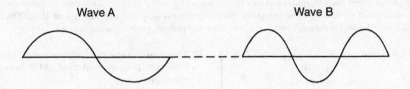

Wave A   Wave B

Which characteristic is the same for both waves?

(1) amplitude   (3) period
(2) frequency   (4) wavelength

35 The diagram below shows a periodic wave.

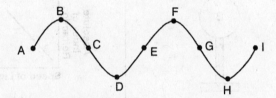

Which two points on the wave are 180.° out of phase?

(1) A and C   (3) F and G
(2) B and E   (4) D and H

Part B–1

Answer all questions in this part.

Directions (36–50): For *each* statement or question, choose the word or expression that, of those given, best completes the statement or answers the question. Some questions may require the use of the *2006 Edition Reference Tables for Physical Setting/Physics*. Record your answers on your separate answer sheet.

36 The height of a 30-story building is approximately
(1) 10^0 m   (3) 10^2 m
(2) 10^1 m   (4) 10^3 m

37 Two identically-sized metal spheres on insulating stands are positioned as shown below. The charge on sphere A is -4.0×10^{-6} coulomb and the charge on sphere B is -8.0×10^{-6} coulomb.

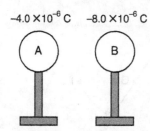

-4.0×10^{-6} C -8.0×10^{-6} C

A   B

The two spheres are touched together and then separated. The total number of excess electrons on sphere A after the separation is
(1) 2.5×10^{13} (3) 5.0×10^{13}
(2) 3.8×10^{13} (4) 7.5×10^{13}

38 A 1.0×10^3-kilogram car travels at a constant speed of 20. meters per second around a horizontal circular track. The diameter of the track is 1.0×10^2 meters. The magnitude of the car's centripetal acceleration is
(1) 0.20 m/s²   (3) 8.0 m/s²
(2) 2.0 m/s²   (4) 4.0 m/s²

39 Which combination of units can be used to express electrical energy?
(1) $\dfrac{\text{volt}}{\text{coulomb}}$

(2) $\dfrac{\text{coulomb}}{\text{volt}}$

(3) volt•coulomb

(4) volt•coulomb•second

40 The total amount of electrical energy used by a 315-watt television during 30.0 minutes of operation is
(1) 5.67×10^5 J   (3) 1.05×10^1 J
(2) 9.45×10^3 J   (4) 1.75×10^{-1} J

41 Which graph best represents the relationship between the absolute index of refraction and the speed of light ($f = 5.09 \times 10^{14}$ Hz) in various media?

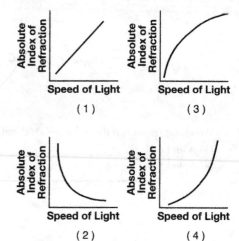

42 A 25-gram paper cup falls from rest off the edge of a tabletop 0.90 meter above the floor. If the cup has 0.20 joule of kinetic energy when it hits the floor, what is the total amount of energy converted into internal (thermal) energy during the cup's fall?
(1) 0.02 J   (3) 2.2 J
(2) 0.22 J   (4) 220 J

43 Which electron transition between the energy levels of hydrogen causes the emission of a photon of visible light?
(1) $n = 6$ to $n = 5$ (3) $n = 5$ to $n = 2$
(2) $n = 5$ to $n = 6$ (4) $n = 2$ to $n = 5$

44 Which graph best represents an object in equilibrium moving in a straight line?

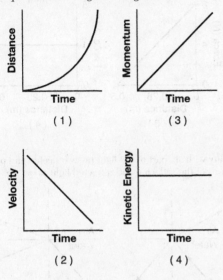

(1) (3)

(2) (4)

45 A body, *B*, is moving at constant speed in a horizontal circular path around point *P*. Which diagram shows the direction of the velocity (v) and the direction of the centripetal force (F_c) acting on the body?

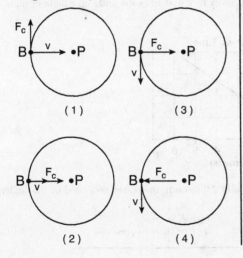

(1) (3)

(2) (4)

46 Which graph best represents the relationship between photon energy and photon wavelength?

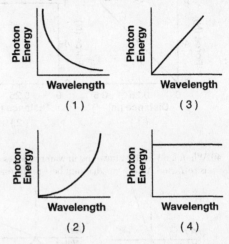

(1) (3)

(2) (4)

47 Which combination of initial horizontal velocity, (v_H) and initial vertical velocity, (v_v) results in the greatest horizontal range for a projectile over level ground? [Neglect friction.]

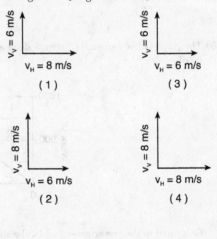

(1) (3)

(2) (4)

48 Which graph best represents the greatest amount of work?

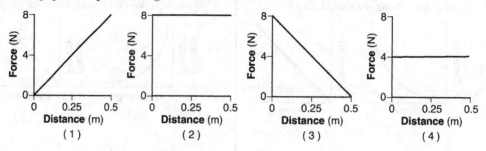

49 When a ray of light traveling in water reaches a boundary with air, part of the light ray is reflected and part is refracted. Which ray diagram best represents the paths of the reflected and refracted light rays?

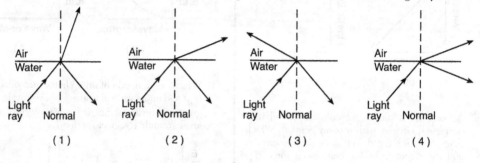

50 The graph below represents the work done against gravity by a student as she walks up a flight of stairs at constant speed.

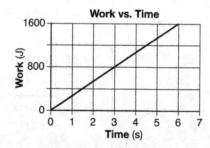

Compared to the power generated by the student after 2.0 seconds, the power generated by the student after 4.0 seconds is

(1) the same   (3) half as great
(2) twice as great   (4) four times as great

Part B–2

Answer all questions in this part.

Directions (51–65): Record your answers in the spaces provided in your answer booklet. Some questions may require the use of the *2006 Edition Reference Tables for Physical Setting/Physics.*

Base your answers to questions 51 through 54 on the information below and the scaled vector diagram in your answer booklet and on your knowledge of physics.

Two forces, a 60.-newton force east and an 80.-newton force north, act concurrently on an object located at point *P*, as shown.

51 Using a ruler, determine the scale used in the vector diagram. [1]

52 Draw the resultant force vector to scale on the diagram *in your answer booklet*. Label the vector "*R*." [1]

53 Determine the magnitude of the resultant force, *R*. [1]

54 Determine the measure of the angle, in degrees, between north and the resultant force, *R*. [1]

55–56 A 3.00-newton force causes a spring to stretch 60.0 centimeters. Calculate the spring constant of this spring. [Show all work, including the equation and substitution with units.] [2]

57 A 7.28-kilogram bowling ball traveling 8.50 meters per second east collides head-on with a 5.45 kilogram bowling ball traveling 10.0 meters per second west. Determine the magnitude of the total momentum of the two-ball system after the collision. [1]

58–59 Calculate the average power required to lift a 490-newton object a vertical distance of 2.0 meters in 10. seconds. [Show all work, including the equation and substitution with units.] [2]

60 The diagram *in your answer booklet* shows wave fronts approaching an opening in a barrier. The size of the opening is approximately equal to one-half the wavelength of the waves. On the diagram *in your answer booklet*, draw the shape of *at least three* of the wave fronts after they have passed through this opening. [1]

61 The diagram *in your answer booklet* shows a mechanical transverse wave traveling to the right in a medium. Point *A* represents a particle in the medium. Draw an arrow originating at point *A* to indicate the initial direction that the particle will move as the wave continues to travel to the right in the medium. [1]

62 Regardless of the method used to generate electrical energy, the amount of energy provided by the source is always greater than the amount of electrical energy produced. Explain why there is a difference between the amount of energy provided by the source and the amount of electrical energy produced. [1]

Base your answers to questions 63 through 65 on the graph below, which represents the relationship between velocity and time for a car moving along a straight line, and your knowledge of physics.

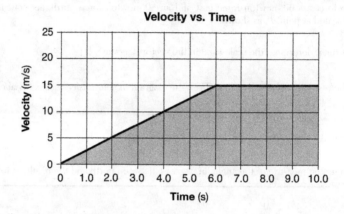

63 Determine the magnitude of the average velocity of the car from $t = 6.0$ seconds to $t = 10$. seconds. [1]

64 Determine the magnitude of the car's acceleration during the first 6.0 seconds. [1]

65 Identify the physical quantity represented by the shaded area on the graph. [1]

Part C
Answer all questions in this part.

Directions (66–85): Record your answers in the spaces provided in your answer booklet. Some questions may require the use of the *2006 Edition Reference Tables for Physical Setting/Physics.*

Base your answers to questions 66 through 70 on the information below and on your knowledge of physics.

A student constructed a series circuit consisting of a 12.0-volt battery, a 10.0-ohm lamp, and a resistor. The circuit does *not* contain a voltmeter or an ammeter. When the circuit is operating, the total current through the circuit is 0.50 ampere.

66 In the space *in your answer booklet*, draw a diagram of the series circuit constructed to operate the lamp, using symbols from the *Reference Tables for Physical Setting/Physics.* [1]

67 Determine the equivalent resistance of the circuit. [1]

68 Determine the resistance of the resistor. [1]

69–70 Calculate the power consumed by the lamp. [Show all work, including the equation and substitution with the units.] [2]

Base your answers to questions 71 through 75 on the information below and on your knowledge of physics.

Pluto orbits the Sun at an average distance of 5.91×10^{12} meters. Pluto's diameter is 2.30×10^6 meters and its mass is 1.31×10^{22} kilograms.

Charon orbits Pluto with their centers separated by a distance of 1.96×10^7 meters. Charon has a diameter of 1.21×10^6 meters and a mass of 1.55×10^{21} kilograms.

71–72 Calculate the magnitude of the gravitational force of attraction that Pluto exerts on Charon. [Show all work, including the equation and substitution with units.] [2]

73–74 Calculate the magnitude of the acceleration of Charon toward Pluto. [Show all work, including the equation and substitution with units.] [2]

75 State the reason why the magnitude of the Sun's gravitational force on Pluto is greater than the magnitude of the Sun's gravitational force on Charon. [1]

Base your answers to questions 76 through 80 on the information below and on your knowledge of physics.

A horizontal 20.-newton force is applied to a 5.0-kilogram box to push it across a rough, horizontal floor at a constant velocity of 3.0 meters per second to the right.

76 Determine the magnitude of the force of friction acting on the box. [1]

77–78 Calculate the weight of the box. [Show all work, including the equation and substitution with units.] [2]

79–80 Calculate the coefficient of kinetic friction between the box and the floor. [Show all work, including the equation and substitution with units] [2]

Base your answers to questions 81 through 85 on the information below and on your knowledge of physics.

An electron traveling with a speed of 2.50×10^6 meters per second collides with a photon having a frequency of 1.00×10^{16} hertz. After the collision, the photon has 3.18×10^{-18} joule of energy.

81–82 Calculate the original kinetic energy of the electron. [Show all work, including the equation and substitution with units.] [2]

83 Determine the energy in joules of the photon before the collision. [1]

84 Determine the energy lost by the photon during the collision. [1]

85 Name *two* physical quantities conserved in the collision. [1]

The University of the State of New York

REGENTS HIGH SCHOOL EXAMINATION

PHYSICAL SETTING
PHYSICS

ANSWER BOOKLET

☐ Male

Student .... Sex: ☐ Female

Teacher ....

School .... Grade ....

Record your answers for Part B–2 and Part C in this booklet.

Part B–2

51 1.0 cm = ___13.6___ N

52

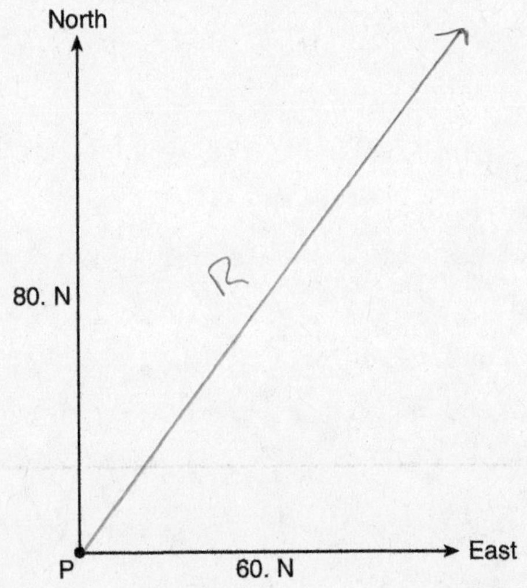

53 ___102___ N

54 ___36___ °

55–56

$$F = Kx$$
$$3 = K(.6)$$
$$K = 5 \frac{N}{m}$$

57 ___7.38___ kg•m/s

58–59

$$P = \frac{Fd}{t}$$
$$P = \frac{490 \cdot 2}{10}$$
$$P = 98 \, \omega$$

60

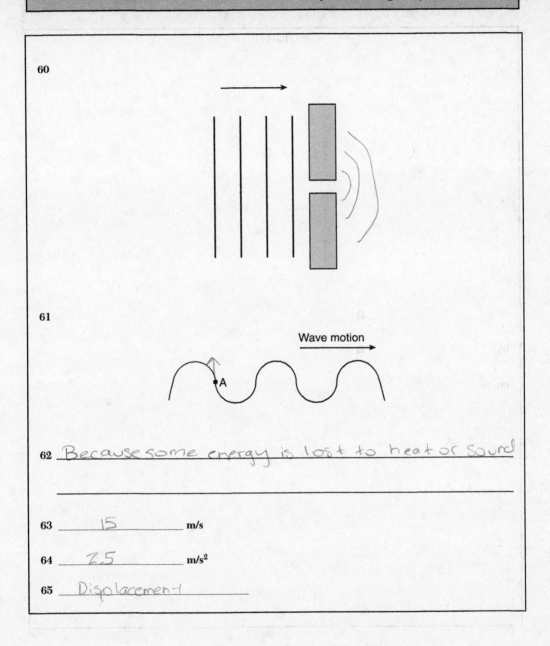

61

Wave motion

A

62 <u>Because some energy is lost to heat or sound</u>

63 <u>15</u> m/s

64 <u>2.5</u> m/s²

65 <u>Displacement</u>

Part C

66

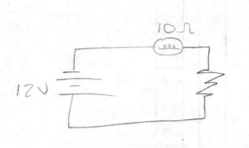

67 _____ 2 4 _____ Ω

68 _____ 1 4 _____ Ω

69–70

$P = I^2 R$

$P = (.5)^2 (10)$

$P = 2.5 W$

71–72

$$F = \frac{G M_1 M_2}{r^2}$$

$$F = \frac{(6.67 \times 10^{-11})(1.55 \times 10^{21})(1.31 \times 10^{22})}{(1.96 \times 10^7)^2}$$

$$F = 3.53 \times 10^{18} N$$

73–74

$$F = ma$$

$$3.53 \times 10^{18} = 1.55 \times 10^{21} a$$

$$a = .0023 \, m/s^2$$

75 Because pluto is more massive than Charon

76 _____20_____ N

77–78

$$F = mg$$
$$F = (5)(9.8)$$
$$F = 49N$$

79–80

$$F_f = \mu mg$$
$$20 = \mu 49$$
$$\mu = .41$$

81–82

$KE = \frac{1}{2}mv^2$

$KE = \frac{1}{2}(2.5\times10^{6})^2 (9.11\times10^{-31})$

$KE = 2.85\times10^{-18}$

83 $\underline{\quad 6.63\times10^{-18} \quad}$ J

84 $\underline{\quad 3.45\times10^{-18} \quad}$ J

85 $\underline{\quad Mass \quad}$ and $\underline{\quad Charge \quad}$

Part A
Answer all questions in this part.

Directions (1–35): For *each* statement or question, choose the word or expression that, of those given, best completes the statement or answers the question. Some questions may require the use of the *2006 Edition Reference Tables for Physical Setting/Physics*. Record your answers on your separate answer sheet.

1 Which quantities are scalar?

   (1) speed and work
   (2) velocity and force
   (3) distance and acceleration
   (4) momentum and power

2 A 3.00-kilogram mass is thrown vertically upward with an initial speed of 9.80 meters per second. What is the maximum height this object will reach? [Neglect friction.]

   (1) 1.00 m   (3) 9.80 m
   (2) 4.90 m   (4) 19.6 m

3 An airplane traveling north at 220. meters per second encounters a 50.0-meters-per-second crosswind from west to east, as represented in the diagram below.

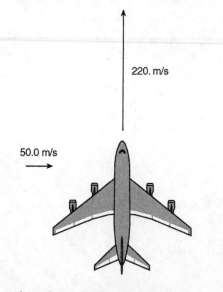

220. m/s

50.0 m/s

What is the resultant speed of the plane?

   (1) 170. m/s   (3) 226 m/s
   (2) 214 m/s   (4) 270. m/s

4 A 160.-kilogram space vehicle is traveling along a straight line at a constant speed of 800. meters per second. The magnitude of the net force on the space vehicle is

   (1) 0 N   (3) 8.00×10^2 N
   (2) 1.60×10^2 N   (4) 1.28×10^5 N

5 A student throws a 5.0-newton ball straight up. What is the net force on the ball at its maximum height?

   (1) 0.0 N   (3) 5.0 N, down
   (2) 5.0 N, up   (4) 9.8 N, down

6 A vertical spring has a spring constant of 100. newtons per meter. When an object is attached to the bottom of the spring, the spring changes from its unstretched length of 0.50 meter to a length of 0.65 meter. The magnitude of the weight of the attached object is

   (1) 1.1 N   (3) 50. N
   (2) 15 N   (4) 65 N

7 A 1.5-kilogram cart initially moves at 2.0 meters per second. It is brought to rest by a constant net force in 0.30 second. What is the magnitude of the net force?

   (1) 0.40 N   (3) 10. N
   (2) 0.90 N   (4) 15 N

8 Which characteristic of a light wave must increase as the light wave passes from glass into air?

   (1) amplitude   (3) period
   (2) frequency   (4) wavelength

9 As a 5.0×10^2-newton basketball player jumps from the floor up toward the basket, the magnitude of the force of her feet on the floor is 1.0×10^3 newtons. As she jumps, the magnitude of the force of the floor on her feet is

(1) 5.0×10^2 N   (3) 1.5×10^3 N

(2) 1.0×10^3 N   (4) 5.0×10^5 N

10 A 0.0600-kilogram ball traveling at 60.0 meters per second hits a concrete wall. What speed must a 0.0100-kilogram bullet have in order to hit the wall with the same magnitude of momentum as the ball?

(1) 3.60 m/s   (3) 360. m/s

(2) 6.00 m/s   (4) 600. m/s

11 The Hubble telescope's orbit is 5.6×10^5 meters above Earth's surface. The telescope has a mass of 1.1×10^4 kilograms. Earth exerts a gravitational force of 9.1×10^4 newtons on the telescope. The magnitude of Earth's gravitational field strength at this location is

(1) 1.5×10^{-20} N/kg   (3) 8.3 N/kg

(2) 0.12 N/kg   (4) 9.8 N/kg

12 When two point charges are a distance d apart, the magnitude of the electrostatic force between them is F. If the distance between the point charges is increased to $3d$, the magnitude of the electrostatic force between the two charges will be

(1) $\frac{1}{9}F$ (3) $2F$

(2) $\frac{1}{3}F$ (4) $4F$

13 A radio operating at 3.0 volts and a constant temperature draws a current of 1.8×10^{-4} ampere. What is the resistance of the radio circuit?

(1) 1.7×10^4 Ω   (3) 5.4×10^{-4} Ω

(2) 3.0×10^1 Ω   (4) 6.0×10^{-5} Ω

14 Which energy transformation occurs in an operating electric motor?

(1) electrical → mechanical

(2) mechanical → electrical

(3) chemical → electrical

(4) electrical → chemical

15 A block slides across a rough, horizontal tabletop. As the block comes to rest, there is an increase in the block-tabletop system's

(1) gravitational potential energy

(2) elastic potential energy

(3) kinetic energy

(4) internal (thermal) energy

16 How much work is required to move an electron through a potential difference of 3.00 volts?

(1) 5.33×10^{-20} J   (3) 3.00 J

(2) 4.80×10^{-19} J   (4) 1.88×10^{19} J

17 During a laboratory experiment, a student finds that at 20° Celsius, a 6.0-meter length of copper wire has a resistance of 1.3 ohms. The cross-sectional area of this wire is

(1) 7.9×10^{-8} m^2 (3) 4.6×10^0 m^2

(2) 1.1×10^{-7} m^2 (4) 1.3×10^7 m^2

18 A net charge of 5.0 coulombs passes a point on a conductor in 0.050 second. The average current is

(1) 8.0×10^{-8} A   (3) 2.5×10^{-1} A

(2) 1.0×10^{-2} A   (4) 1.0×10^2 A

19 If several resistors are connected in series in an electric circuit, the potential difference across each resistor

(1) varies directly with its resistance

(2) varies inversely with its resistance

(3) varies inversely with the square of its resistance

(4) is independent of its resistance

20 The amplitude of a sound wave is most closely related to the sound's

(1) speed   (3) loudness

(2) wavelength   (4) pitch

21 A duck floating on a lake oscillates up and down 5.0 times during a 10.-second interval as a periodic wave passes by. What is the frequency of the duck's oscillations?

(1) 0.10 Hz   (3) 2.0 Hz

(2) 0.50 Hz   (4) 50. Hz

22 Which diagram best represents the position of a ball, at equal time intervals, as it falls freely from rest near Earth's surface?

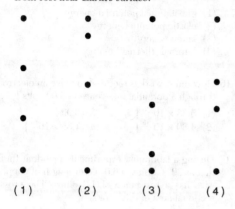

(1)   (2)   (3)   (4)

23 A gamma ray and a microwave traveling in a vacuum have the same

(1) frequency   (3) speed
(2) period   (4) wavelength

24 A student produces a wave in a long spring by vibrating its end. As the frequency of the vibration is doubled, the wavelength in the spring is

(1) quartered   (3) unchanged
(2) halved   (4) doubled

25 Which two points on the wave shown in the diagram below are in phase with each other?

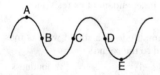

(1) A and B   (3) B and C
(2) A and E   (4) B and D

26 As a longitudinal wave moves through a medium, the particles of the medium

(1) vibrate parallel to the direction of the wave's propagation
(2) vibrate perpendicular to the direction of the wave's propagation
(3) are transferred in the direction of the wave's motion, only
(4) are stationary

27 Wind blowing across suspended power lines may cause the power lines to vibrate at their natural frequency. This often produces audible sound waves. This phenomenon, often called an Aeolian harp, is an example of

(1) diffraction   (3) refraction
(2) the Doppler effect   (4) resonance

28 A student listens to music from a speaker in an adjoining room, as represented in the diagram below.

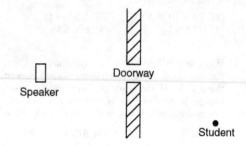

She notices that she does not have to be directly in front of the doorway to hear the music. This spreading of sound waves beyond the doorway is an example of

(1) the Doppler effect   (3) refraction
(2) resonance   (4) diffraction

29 What is the minimum energy required to ionize a hydrogen atom in the $n = 3$ state?

(1) 0.00 eV   (3) 1.51 eV
(2) 0.66 eV   (4) 12.09 eV

Base your answers to questions 30 and 31 on the diagram below and on your knowledge of physics. The diagram represents two small, charged, identical metal spheres, A and B that are separated by a distance of 2.0 meters.

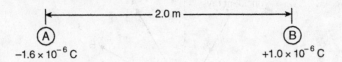

$$|\leftarrow \text{---------} 2.0\,\text{m} \text{---------} \rightarrow|$$

(A) (B)

$-1.6 \times 10^{-6}\,\text{C}$ $+1.0 \times 10^{-6}\,\text{C}$

30 What is the magnitude of the electrostatic force exerted by sphere A on sphere B?

(1) $7.2 \times 10^{-3}\,\text{N}$ (3) $8.0 \times 10^{-13}\,\text{N}$
(2) $3.6 \times 10^{-3}\,\text{N}$ (4) $4.0 \times 10^{-13}\,\text{N}$

31 If the two spheres were touched together and then separated, the charge on sphere A would be

(1) $-3.0 \times 10^{-7}\,\text{C}$ (3) $-1.3 \times 10^{-6}\,\text{C}$
(2) $-6.0 \times 10^{-7}\,\text{C}$ (4) $-2.6 \times 10^{-6}\,\text{C}$

---

32 The horn of a moving vehicle produces a sound of constant frequency. Two stationary observers, A and C, and the vehicle's driver, B, positioned as represented in the diagram below, hear the sound of the horn.

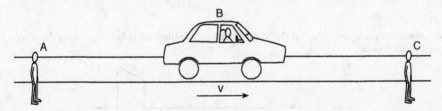

Compared to the frequency of the sound of the horn heard by driver B, the frequency heard by observer A is

(1) lower and the frequency heard by observer C is lower
(2) lower and the frequency heard by observer C is higher
(3) higher and the frequency heard by observer C is lower
(4) higher and the frequency heard by observer C is higher

33 A different force is applied to each of four different blocks on a frictionless, horizontal surface. In which diagram does the block have the greatest inertia 2.0 seconds after starting from rest?

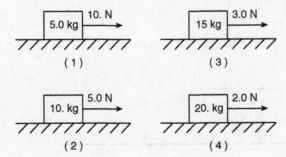

10. N → 5.0 kg	3.0 N → 15 kg
(1)	(3)
5.0 N → 10. kg	2.0 N → 20. kg
(2)	(4)

34 The diagram below shows a ray of monochromatic light incident on a boundary between air and glass.

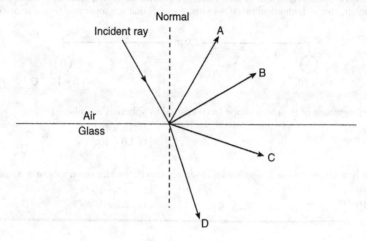

Which ray best represents the path of the reflected light ray?

(1) A   (3) C
(2) B   (4) D

35 Two pulses approach each other in the same medium. The diagram below represents the displacements caused by each pulse.

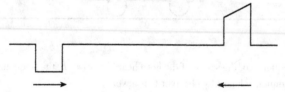

Which diagram best represents the resultant displacement of the medium as the pulses pass through each other?

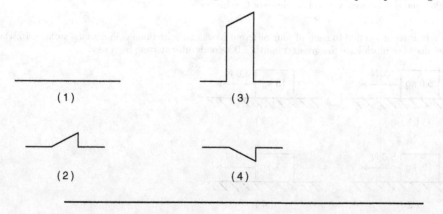

(1)   (3)

(2)   (4)

Part B–1

Answer all questions in this part.

Directions (36–50): For *each* statement or question, choose the word or expression that, of those given, best completes the statement or answers the question. Some questions may require the use of the *2006 Edition Reference Tables for Physical Setting/Physics*. Record your answers on your separate answer sheet.

36 The diameter of an automobile tire is closest to

(1) 10^{-2} m   (3) 10^1 m

(2) 10^0 m   (4) 10^2 m

37 The vector diagram below represents the velocity of a car traveling 24 meters per second 35° east of north.

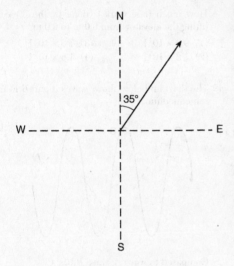

What is the magnitude of the component of the car's velocity that is directed eastward?

(1) 14 m/s   (3) 29 m/s

(2) 20. m/s   (4) 42 m/s

38 Without air resistance, a kicked ball would reach a maximum height of 6.7 meters and land 38 meters away. With air resistance, the ball would travel

(1) 6.7 m vertically and more than 38 m horizontally

(2) 38 m horizontally and less than 6.7 m vertically

(3) more than 6.7 m vertically and less than 38 m horizontally

(4) less than 38 m horizontally and less than 6.7 m vertically

39 A car is moving with a constant speed of 20. meters per second. What total distance does the car travel in 2.0 minutes?

(1) 10. m   (3) 1200 m

(2) 40. m   (4) 2400 m

40 A car, initially traveling at 15 meters per second north, accelerates to 25 meters per second north in 4.0 seconds. The magnitude of the average acceleration is

(1) 2.5 m/s^2 (3) 10. m/s^2

(2) 6.3 m/s^2 (4) 20. m/s^2

41 An object is in equilibrium. Which force vector diagram could represent the force(s) acting on the object?

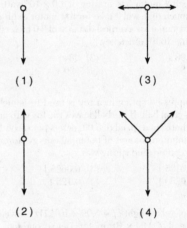

42 Which combination of fundamental units can be used to express the amount of work done on an object?

(1) kg•m/s   (3) kg•m^2/s^2

(2) kg•m/s^2 (4) kg•m^2/s^3

43 Which graph best represents the relationship between the potential energy stored in a spring and the change in the spring's length from its equilibrium position?

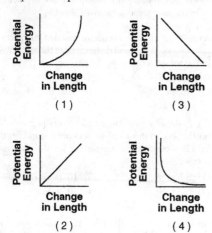

(1)

(3)

(2)

(4)

44 An electric motor has a rating of 4.0×10^2 watts. How much time will it take for this motor to lift a 50.-kilogram mass a vertical distance of 8.0 meters? [Assume 100% efficiency.]

(1) 0.98 s (3) 98 s
(2) 9.8 s   (4) 980 s

45 A compressed spring in a toy is used to launch a 5.00-gram ball. If the ball leaves the toy with an initial horizontal speed of 5.00 meters per second, the minimum amount of potential energy stored in the compressed spring was

(1) 0.0125 J (3) 0.0625 J
(2) 0.0250 J (4) 0.125 J

46 A ray of yellow light ($f = 5.09 \times 10^{14}$ Hz) travels at a speed of 2.04×10^8 meters per second in

(1) ethyl alcohol (3) Lucite
(2) water   (4) glycerol

47 A blue-light photon has a wavelength of 4.80×10^{-7} meter. What is the energy of the photon?

(1) 1.86×10^{22} J (3) 4.14×10^{-19} J
(2) 1.44×10^2 J   (4) 3.18×10^{-26} J

48 The graph below represents the relationship between the force exerted on an elevator and the distance the elevator is lifted.

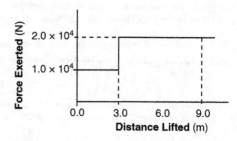

How much total work is done by the force in lifting the elevator from 0.0 m to 9.0 m?

(1) 9.0×10^4 J (3) 1.5×10^5 J
(2) 1.2×10^5 J (4) 1.8×10^5 J

49 The diagram below shows waves A and B in the same medium.

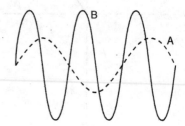

Compared to wave A, wave B has

(1) twice the amplitude and twice the wavelength
(2) twice the amplitude and half the wavelength
(3) the same amplitude and half the wavelength
(4) half the amplitude and the same wavelength

50 What is the quark composition of a proton?

(1) uud (3) csb
(2) udd (4) uds

Part B–2

Answer all questions in this part.

Directions (51–65): Record your answers in the spaces provided in your answer booklet. Some questions may require the use of the *2006 Edition Reference Tables for Physical Setting/Physics*.

51–52 Calculate the minimum power output of an electric motor that lifts a 1.30×10^4-newton elevator car vertically upward at a constant speed of 1.50 meters per second. [Show all work, including the equation and substitution with units.] [2]

53–54 A microwave oven emits a microwave with a wavelength of 2.00×10^{-2} meter in air. Calculate the frequency of the microwave. [Show all work, including the equation and substitution with units.] [2]

55–56 Calculate the energy equivalent in joules of the mass of a proton. [Show all work, including the equation and substitution with units.] [2]

Base your answers to questions 57 through 59 on the information and diagram below and on your knowledge of physics.

A 1.5×10^3-kilogram car is driven at a constant speed of 12 meters per second counterclockwise around a horizontal circular track having a radius of 50. meters, as represented below.

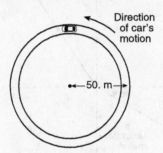

Direction of car's motion

◄—50. m—►

Track, as Viewed from Above

57 On the diagram *in your answer booklet*, draw an arrow to indicate the direction of the velocity of the car when it is at the position shown. Start the arrow on the car. [1]

58–59 Calculate the magnitude of the centripetal acceleration of the car. [Show all work, including the equation and substitution with units.] [2]

Base your answers to questions 60 through 62 on the information below and on your knowledge of physics.

A football is thrown at an angle of 30.° above the horizontal. The magnitude of the horizontal component of the ball's initial velocity is 13.0 meters per second. The magnitude of the vertical component of the ball's initial velocity is 7.5 meters per second. [Neglect friction.]

60 On the axes *in your answer booklet*, draw a graph representing the relationship between the horizontal displacement of the football and the time the football is in the air. [1]

61–62 The football is caught at the same height from which it is thrown. Calculate the total time the football was in the air. [Show all work, including the equation and substitution with units.] [2]

_____

Base your answers to questions 63 through 65 on the information and diagram below and on your knowledge of physics.

A ray of light ($f = 5.09 \times 10^{14}$ Hz) traveling through a block of an unknown material, passes at an angle of incidence of 30.° into air, as shown in the diagram below.

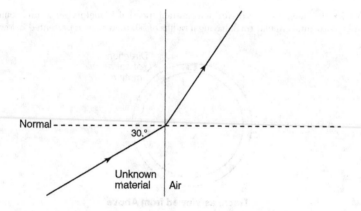

63 Use a protractor to determine the angle of refraction of the light ray as it passes from the unknown material into air. [1]

64–65 Calculate the index of refraction of the unknown material. [Show all work, including the equation and substitution with units.] [2]

_____

Part C

Answer all questions in this part.

Directions (66–85): Record your answers in the spaces provided in your answer booklet. Some questions may require the use of the *2006 Edition Reference Tables for Physical Setting/Physics.*

Base your answers to questions 66 through 70 on the information below and on your knowledge of physics.

The diagram below represents a 4.0-newton force applied to a 0.200-kilogram copper block sliding to the right on a horizontal steel table.

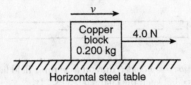

66 Determine the weight of the block. [1]

67–68 Calculate the magnitude of the force of friction acting on the moving block. [Show all work, including the equation and substitution with units.] [2]

69 Determine the magnitude of the net force acting on the moving block. [1]

70 Describe what happens to the magnitude of the velocity of the block as the block slides across the table. [1]

Base your answers to questions 71 through 75 on the information and diagram below and on your knowledge of physics.

Two conducting parallel plates 5.0×10^{-3} meter apart are charged with a 12-volt potential difference. An electron is located midway between the plates. The magnitude of the electrostatic force on the electron is 3.8×10^{-16} newton.

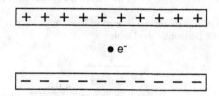

71 On the diagram *in your answer booklet*, draw *at least three* field lines to represent the direction of the electric field in the space between the charged plates. [1]

72 Identify the direction of the electrostatic force that the electric field exerts on the electron. [1]

73–74 Calculate the magnitude of the electric field strength between the plates, in newtons per coulomb. [Show all work, including the equation and substitution with units.] [2]

75 Describe what happens to the magnitude of the net electrostatic force on the electron as the electron is moved toward the positive plate. [1]

Base your answers to questions 76 through 80 on the information below and on your knowledge of physics.

An electron in a mercury atom changes from energy level b to a higher energy level when the atom absorbs a single photon with an energy of 3.06 electronvolts.

76 Determine the letter that identifies the energy level to which the electron jumped when the mercury atom absorbed the photon. [1]

77 Determine the energy of the photon, in joules. [1]

78–79 Calculate the frequency of the photon. [Show all work, including the equation and substitution with units.] [2]

80 Classify the photon as one of the types of electromagnetic radiation listed in the electromagnetic spectrum. [1]

Base your answers to questions 81 through 85 on the information and circuit diagram below and on your knowledge of physics.

Three lamps are connected in parallel to a 120.-volt source of potential difference, as represented below.

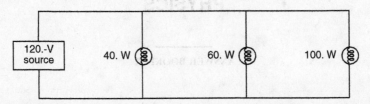

81–82 Calculate the resistance of the 40.-watt lamp. [Show all work, including the equation and substitution with units.]   [2]

83 Describe what change, if any, would occur in the power dissipated by the 100.-watt lamp if the 60.-watt lamp were to burn out.   [1]

84 Describe what change, if any, would occur in the equivalent resistance of the circuit if the 60.-watt lamp were to burn out.   [1]

85 The circuit is disassembled. The same three lamps are then connected in series with each other and the source. Compare the equivalent resistance of this series circuit to the equivalent resistance of the parallel circuit.   [1]

The University of the State of New York

REGENTS HIGH SCHOOL EXAMINATION

PHYSICAL SETTING
PHYSICS

―――――

ANSWER BOOKLET

Student . . . .   Sex: ☐ Male
☐ Female

Teacher . . . .

School . . . .   Grade . . . .

Record your answers for Part B–2 and Part C in this booklet.

Part B–2

51–52

53–54

55–56

57

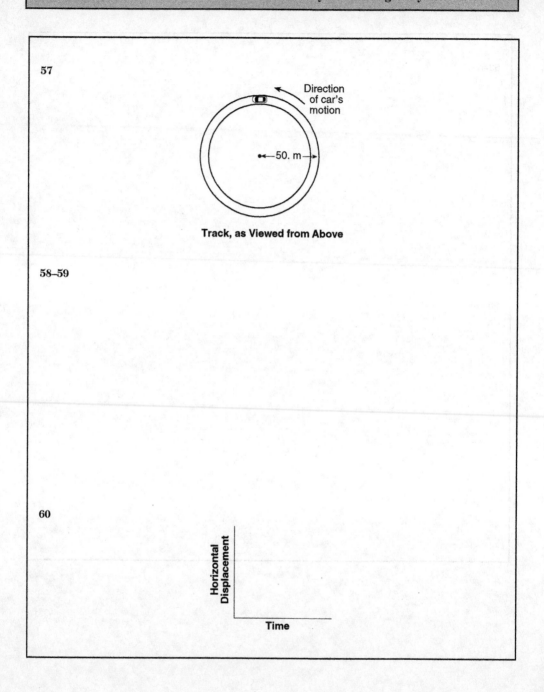

Direction of car's motion

←— 50. m —→

Track, as Viewed from Above

58–59

60

Horizontal Displacement

Time

61–62

63 _____ °

64–65

Part C

66 _____ N

67–68

69 _____ N

70 _____

   _____

71

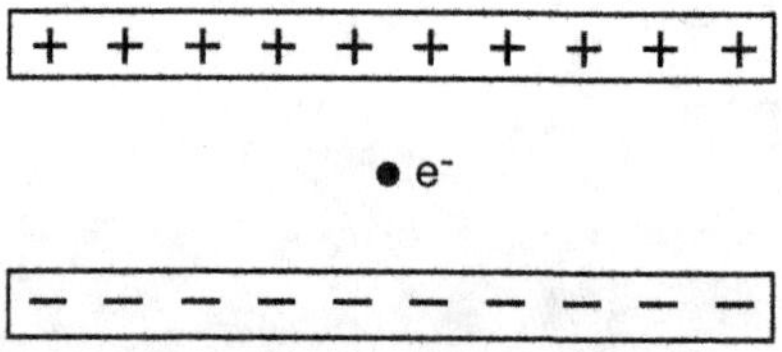

+ + + + + + + + + +

● e⁻

− − − − − − − − − −

72 _____

73–74

75 _____

_____

76 _____

77 _____ **J**

78–79

80 _____

81–82

83 _____

_____

84 _____

_____

85 _____

_____